HITLER'S ESCAPE

History is no more than the portrayal of crimes and misfortunes.

Voltaire

HITLER'S ESCAPE

Ron T. Hansig

ATHENA PRESS
LONDON

ISBN 1 932077 82 0

First Published 2005 by
ATHENA PRESS
Queen's House, 2 Holly Road
Twickenham TW1 4EG
United Kingdom

Printed for Athena Press

Contents

INTRODUCTION AND
REFLECTIONS

When considering the fate of the German Dictator Adolf Hitler, one has to realize that the overwhelming majority of my readers believe that he committed suicide during the last days of the Second World War. This view is supported by dozens of books written by among others, world-renowned historians. Their views and conclusions are seemingly well supported by circumstances and eyewitness accounts. Just the consideration that there may be another, darker side to the story makes us, to say it mildly, emotionally uneasy. Just the thought that such a man, responsible for the murder of millions, might have escaped unscathed from the rubble of Berlin in 1945 seems hard to swallow. Yet, as we shall see, there is sufficient evidence in this book, some from recent sources, to at least consider the possibility that Hitler, together with Eva Braun, fled from Berlin on April 22, 1945.

One could easily dismiss such evidence as part of yet another conspiracy theory. But it serves the interests of free people to openly debate even what most would consider, *well*-established *facts*, despite the emotional problems that this may cause.

There are many unresolved mysteries in human history. One of the more recent ones is the Kennedy assassination on November 22, 1963. Despite all the *well-established facts* surrounding this case, about seventy percent[1] of the U.S. population don't believe that Oswald was the sole killer of the President. Neither did a Congressional Committee.

While there is an *official* version of the Kennedy murder, there is also an official story widely accepted, at least by the Western Allies, that Hitler committed suicide on April 30, 1945. Yet, as with the Kennedy story, there remain too many contradictions and unexplained details that cast doubt on the accuracy of the official version. The Russian[2] dictator, Marshal Stalin, who was in a position to know since his troops occupied Hitler's bunker, told U.S. President Truman and other Western leaders in July of 1945 that Hitler escaped. This clear and unequivocal statement is typically dismissed as "cold war" propaganda. Yet he made these statements only two months after the end of the war against Germany and at the Potsdam Conference, where all three major Western Allies met harmoniously to "divide the spoils" of war. They were still allies in the fight against Japan. The term "cold war" was only coined a year later,[3] in 1946, when tensions between Russia and the West started to build. One of Stalin's successors, Premier Krushchev, stated in his memoirs: "Stalin naturally insisted for a long time that Hitler was not dead at all…" While we may doubt Stalin's words addressed to the Western leaders, the question still remains: what motives would he have to lie to his own closest politburo members?

Stalin's statements should carry a lot of weight since his troops conquered Berlin and occupied Hitler's bunker. His soldiers were the first eyewitnesses at least to the aftermath of the alleged suicide. He, at that time, had the best espionage service in the world and had a "mole" within Hitler's staff! His judgment as to the events in the Hitler bunker therefore carries more weight than that of his allies who had to rely on hearsay and on dubious eyewitness accounts. An example of this was Hitler's chauffeur, E. Kempka, the sole key bunker survivor caught by the West. In 1973 he said, "Back in 1945 I told my interrogators anything they wanted to hear." So much for reliability! The real question is, why did the Western leaders not follow up on Stalin's lead and request Hitler's extradition from Spain? I believe the answer lies in the fact that the Allies needed closure on this subject. Only a dead Hitler could bring an end to "Hitlerism" in Germany. Remember, the British and American Governments started the daunting "Umerziehungsprogramm" (re-indoctrination program) for the German people in order to eradicate Nazism. It is probably for this reason that, according to the *Atlanta Constitution* of May 2, 1945: "The British Foreign Office Believes (Hitler's) Death Report". This of course was done without any proof and without an identified corpse. To later publicly request extradition of Hitler from Spain would be to admit that Hitler was still alive;

not only an embarrassment for the Western Allies, but possibly might have stirred up a rebellion in Germany. It is, of course quite possible that Hitler left Spain and traveled by ship from Barcelona on to Argentina (see Stalin's statements). It is well known, that in 1945 Argentina was a haven to quite a few escaping Nazis, Eichmann and Mengele among them.

After reading several books about the last days of the Second World War my own doubts became stronger and I decided to make a more serious study of the matter, employing a logical approach that a policeman would have while sifting through evidence.

Then again, any author voicing views contrary to established wisdom has to beware – he or she just might be branded a crank or worse.

One may ask, why this morbid interest in one of the most sinister personages of world history? The answer is exactly that. In order to learn from history, we first have to understand it.

Distorted or falsified history may serve a short-term political or propaganda purpose, but in the end it only creates confusion and skepticism.

The aim of this study certainly is not to glorify Hitler, or to make him out as a latter day hero, but to show him as a coward, escaping justice. History provides ample proof of the incredible death and destruction he caused to the Jews, to Germany and to the rest of Europe and Russia.

There is of course the danger that by exposing his escape, that this may tend to add status to the "Fuehrer" by showing him as "thumbing his nose at his enemies", so to speak. While this may be true, we also have to recognize that more than two generations have passed since this happened and we can also be sure that all participants in these crimes are now dead, forgotten or punished. Finally, there should be some recognition of the fate of the unfortunate Herr Sillip, the double of Hitler, who after all was murdered, as we shall see.

Let's realize too, this is also a great detective story, full of hidden clues, false leads, politically motivated forged evidence, at least one apparent murder victim, and a whole bunch of suspicious characters, braggarts and conspirators.

Like any good detective yarn, we have to look for "motive", "opportunity", and "means", in order to resolve this story. As to motive, we may assume Hitler had a will to survive and a resolve for his body not to become an object of public display and ridicule.[4] He may have anticipated what would happen to Mussolini later on April 29, 1945, whose body was strung upside down from the roof of a garage. These are very strong motives indeed. That Hitler was averse to suicide was testified to by the German Field Marshall, Gerd von Rundstedt, who told reporters after the war that: "...never, never will I believe, he [Hitler] put an end to his own life. That was not in accordance with his nature." The Grand-Admiral Doenitz stated: "Given my appreciation of his personality, I did not believe a suicide to be possible..." echoing this sentiment.

As to opportunity, there were still a number of airplanes available on April 22 and 23. The last flight was recorded to have left Berlin as late as April 29, 1945. Also, on April 22, there were still roads open out of Berlin since the city was not surrounded until the afternoon of April 25, 1945.

What was more difficult, as in a "perfect murder", was to leave no trace. Here we come to the means. Hitler had a number of immensely dedicated, fanatical followers to help him to disappear. He had plenty of money and still in April 1945, a well-oiled secret police organization. Just to show how much money was left in Germany, it was reported[5] that U.S. troops discovered, in April of 1945, a total of 8, 198 gold bars each weighing 35 lbs. (Current street value about $4 billion) plus 2, 474 bags of gold coins, several millions of U.S. dollar banknotes, plus precious stones and other valuables in the Kaiserrode salt mines near Eisenach in Saxony.

All indications are that Hitler and Eva Braun, rather than committing suicide, flew to Barcelona, Spain, probably at 8 P.M. on April 26, landing there on April 27, 1945. What nobody apparently has learned, as of this date, is what happened to Hitler afterwards.

I, and surely many others, would be eager to find out. There is the possibility that Franco, then the Spanish leader, put the Hitler party into a guarded safe house, probably in a remote area of the country. However, this lifestyle of enforced idleness must have been unbearable for a person who, in the past, could command millions. On the other hand, the Spanish secret police may have liquidated Hitler quietly, in order to avoid an international embarrassment for Spain. In any case, the man is dead, one way or another, and can no longer serve as a rallying point for neo-Nazi elements. That he escaped just punishment for his many crimes is certainly regrettable. Wherever he or Eva Braun traveled, they certainly did so under false names and false passports. We may therefore never know in which country or how, they finally expired.

At this point one may wonder why did Generalissimo Franco grant Hitler asylum?

This involved substantial political risk for Spain. It is true that Franco owed Hitler a great debt. Without German airplanes establishing air supremacy over the Spanish battlefields the Spanish Civil War might not have been won. It is also true that Franco refused to join Germany in the war against Great Britain. There were several reasons for this. Spain at that time was an exhausted and devastated country, whose army had neither heavy weapons nor an effective air force. Yet despite this refusal, Franco maintained very friendly relations with Germany throughout the war, which included supply of badly needed raw material and allowing a German spy network to operate on Spanish soil. He even supplied one division of Spanish troops to fight with the Germans against Russia.

Finally there were rumors that substantial amounts of gold were mysteriously transferred from the Bank of France to Spain a few days before the Allies liberated Paris.

Admittedly, piecing the puzzle together is very difficult, not only because of the many conflicting stories by witnesses but also because of the conflicting dates. I admit it would be difficult for me to remember where I was five or ten years ago, or what exactly happened on that day. I therefore relied on dates independently given by several sources or persons. As to the stories of eyewitnesses I again looked for multiple verification; if that was not possible, I tried to apply logic and common sense. A lot of so-called witnesses tend to embellish their tales. An example would be Dr. Schenck, who wandered into the bunker one day before the assumed "suicide of Hitler" and who seems to have styled himself as an expert witness. Yet his reliability is sometimes questionable. For example, he quotes Ambassador Hewel of telling him that he (Hewel) gave Hitler every morning at 7.30 A.M. a breakfast briefing on foreign affairs. This cannot be true, since Hitler usually got up around noon due to his late night working habits. On the other hand, the Hitler double did wake up early in the morning, the only "Fuehrer" that Schenck observed, if only for a day.

One important clue about Hitler's likely disappearance lies in the differences in his physical appearance, mental health and general demeanor both before and after his likely escape on April 22, 1945. Many historians failed to consider these distinctions, but it provided me with valuable clues for this investigation. Hitler certainly was very fatigued and tired during the last months of the war. However, this seemed not to have affected his mental and intellectual capabilities up to the time when he left Berlin. As to his physical condition, as the last newsreels and still photos taken on April 20, 1945, attest, he certainly was not the "human wreck", hardly able to shuffle, a description that was applied to the person pretending to be Hitler after April 22, 1945.

Here are the main points that made me decide to investigate this matter:

1. Marshal Stalin's insistence that Hitler fled to Spain or Argentina.
2. The fact that the Russians could not produce a corpse of Hitler.
3. The statement by SS General Mueller that Hitler flew to Spain on April 26, 1945.
4. The fact that the Russians found the buried and unburned corpse of Hitler's double close to the bunker.

Unfortunately, the late date eliminates all chances of finding eyewitnesses that may still be alive. The real story of what happened went with them to their graves. Yet there is sufficient new evidence that I discovered, partly in unearthed documentation and in recently published books on the subject, to at least throw serious doubt on the established version of Hitler's death.

I have to leave it up to the readers to decide what really happened in Berlin.

Notes

[1] From a November 2003 CNN television survey.

[2] I typically use the word "Russian" instead of the former "Soviet", and Russian Army for "Red Army", in order to make it easier for the modern reader.

[3] By Herbert Bayard Swope.

[4] He expressed this fear many times to his sub-ordinates.

[5] Brown, Anthony Cave, *The Last Hero, Wild Bill Donovan*, Vintage Books, a division of Random House, 1984

WHO IS THIS HITLER?

I thought it best to make my readers, especially the younger ones, familiar with the historical background and to provide the setting as it were to the climax of World War Two in the spring of 1945. It is now almost sixty years ago when this cataclysm occurred and memories do fade. What we learn about Hitler is usually restricted to reruns of old movies about him, or television entertainment masquerading as historical documentation.

History books in themselves do not always provide a clear and unbiased view or reference source either, especially when the subject of the book is publicly despised.

As an example, we are hard pressed to find an objective appraisal of Attila the Hun. Even though he died over1500 years ago, his name still sends shudders down our spines; yet few know that he was, for his age, a well-educated man and a good administrator of an empire stretching at one time from the Asian steppes to the center of today's France.

To understand Hitler, one has to understand the recent history of Germany, since both are intertwined. When World War One ended on November 11, 1918, Germany was a defeated and exhausted nation even though no foreign troops invaded its soil. However, the German Emperor was forced to abdicate and there was a revolution going on. While the revolutionaries were mainly socialist (a group of less radical adherers to Karl Marx's teachings), the trappings were all the same as what happened in 1917 in St. Petersburg, Russia, with soldiers wearing red arm bands roamed the streets and gangs of civilians took over local and state governments. It was only with the help of loyal elements of the Army that some order was restored and the so-called Weimar Republic was formed. Economic conditions, partly the results of high war reparation forced upon Germany by the victorious Allies, and later the economic depression of the late 1920s resulted in a very high rate of unemployment. In addition, the

1920s saw hyper-inflation reducing the exchange rate from the original 4 marks per dollar to 130 million marks per dollar! These conditions created even more unrest and civil strife, which, at times, bordered on anarchy. It is against this background that Hitler was eventually able to come to power, being democratically elected, in 1933.

Who was this Hitler? He was not even German, being born on April 20, 1889 in Braunau, then part of the Austro-Hungarian Empire. His father was a customs official and rumored to have been the illegitimate son of the Jewish employer of Hitler's grandmother. Hitler disliked his strict father and favored his mother. As a young boy he read a lot, a habit that he retained later on and which provided him with most of his, sometimes encyclopedic, but patchy knowledge. He quit high school in 1906 and he spent his time going to theaters and operas in the provincial capital of Linz. He also started to copy romantic paintings. In 1907, when he was eighteen years old, his mother became incurably ill with cancer and Hitler left home to study to become a painter in Vienna.

Even though his application to the Academy of Art was rejected, he probably taught himself how to paint, since he painted and sold quite a few paintings, mostly watercolors, primarily depicting buildings and churches. He had a great interest in architecture. Yet he already read a lot about racial patterns in people (the Eugenics Movement, which was then in scientific vogue) and nationalistic literature, which, in part molded his later political outlook.

In 1913 he moved to Munich, perhaps to escape being drafted into the Austrian Army.

When World War One started in 1914, he volunteered and joined the German Army. Here he served as a dispatch runner on the Western Front, he was wounded twice, and advanced to Corporal and received the Iron Cross, First and Second Class, for conspicuous bravery.

He learned of the armistice while recuperating in a hospital. This news filled him with bitterness. After discharge he moved to Munich where he worked as a police intelligence agent, infiltrating the different local parties. One of the parties, The

German Workers' Party, appealed to him and he joined. He was member number fifty-five but soon rose to become a member of their executive committee. He then changed the name of the group to: National Socialist German Workers' Party (abbreviated in German as NSDAP).

In 1923 he sensed it might be time to march on Berlin and to replace the "Jewish-Marxist Traitors", as he called the government, but first his party attempted to take over the Bavarian State Government. He tried to do this by marching with his followers to the government buildings, trying to occupy them by force. However this attempt, on November 8, 1923, was blocked by gunfire from police, and several party members were killed or wounded. Hitler subsequently was arrested for treason, was convicted and sentenced to five years' imprisonment. While in prison at Landsberg he wrote his book, *Mein Kampf*, in which he espoused his political philosophies.

After his early release in December 1924, he settled in Berchtesgaden to plan the future, deciding for example to forego revolution and instead prepare to gain power through elections. Having already come to public attention, he now met with such important people as Neville Chamberlain of England and Kurt von Schuschnigg, the Premier of Austria. Part of this attention was the search by the Western Powers, especially England, to find within Germany a party that could be powerful enough to block the rise of the Communist party (which by then was already one of the largest single parties in Germany). In 1919, the Communists already occupied and ran the state of Bavaria, for a few months, till their leader Reisener was assassinated by Count Arco Valti, and a volunteer army drove them out. As a result, there was a universal fear (shared by Winston Churchill) that the whole of Germany might be taken over by the Communists who, in turn, then might spread over Western Europe. This is the reason for Hitler's early support, financial and otherwise, by foreign and domestic businessmen and politicians alike. It is revealing that in 1930 even Winston Churchill wrote:

> ...that authoritarian leaders might be a new and salutary alternative to the weakening, inefficient and increasingly unrepresentative parliamentary systems in many parts of Europe.[1]

While Hitler may have been despised by many in private, his party and his philosophy was considered the lesser of two evils. After all, he would not expropriate private property nor eliminate wage and income differentials, as the Communists promised. His nationalism appealed to the broad middle-class masses, especially after what was considered the unfair treatment of Germany at Versailles. As a result, he rapidly expanded his power and the support for him and his party.

He was a charismatic speaker, who was not only a master of oratory, but he planned his speeches to include emotional punch lines. Later on he had his rallies brilliantly stage-managed. In addition, he had a lot of energy. On one occasion he gave speeches in twenty-one cities within one week! He mesmerized his audience, especially women, some of whom exhibited the near hysteria that we later found with rock band groupies and the deaths of Eva Peron and Princess Diane. Hitler also appealed to the lower middle classes by promising them jobs, and to the military by telling them he would defy the Treaty of Versailles and build a great army. Finally, he promised leaders of industry to begin a rearmament and construction program if he was elected.

All this effort paid off. While in the Federal elections of 1928 his party only got 7.6% of the seats in the Reichstag (Parliament) compared to a combined total of 40.4% for Socialist and Communist, he was able to increase his party's seats to 33.1% during the elections in November of 1932. This made him the leader of the largest faction and, after forming an alliance with some smaller conservative parties, gave him a voting majority. As a result he was asked by then President von Hindenburg to form a new Government, with Hitler appointed to the post of Chancellor.

In retrospect, one has to consider the tumultuous times. Six million people were unemployed; there were almost daily street battles between the competing parties, since each of the major parties had their own uniformed paramilitary organizations. For example, the Communists had their "Red Front Fighting Units" while the Nazis had their SA or "Storm Troopers", the Socialists had yet another outfit, and so on. Just to show how severe this street fighting was, it was reported on May 1, 1928 during fighting between the Communist Red Front and the police, that there were 31 people killed.[2]

The new Government, taking office on January 30, 1933, and led by Hitler, proved to be very energetic. It restored law and order and in September began the construction of a network of autobahns. These and other large construction projects drove the unemployment down from six to less than two million workers in less than six months! In addition, he launched the production of a low-cost radio and later of the famous Volkswagen car. This was, of course, very welcome news to an impoverished population that lost nearly everything during the First World War, then again during the hyperinflation of 1923, and finally during the world wide depression. As a result of these positive measures Hitler's mandate increased from 33.1% in November 1932 to 45% in March of 1933, just four months later. The majority of the German people gladly accepted the positive achievements and overlooked the shadows and the more sinister side of Hitler's activities.

In his first year in office, he dissolved the Communist party and had their leaders interned in the newly established concentration camp near Dachau in Bavaria, one of many to follow. Later in 1933 he called for a boycott of Jewish goods and stores, expulsion of Jews from Germany and, during the war, by open persecution and murder of Jews throughout Europe.

What kind of person was Hitler? He certainly had a brilliant mind, and he was completely ruthless. For example, in 1934, he eliminated the leadership of the SA (Storm Troopers), who tried to create a second Army in Germany and threatened Hitler's support of the armed forces. He admired Stalin for his ruthlessly extermination of all of his opponents. In a rare self-analysis during the last weeks of April 1945, he regretted that he had not been ruthless enough. He also admitted to serious political blunders, such has his alliance with the Italian dictator Mussolini. On the other hand, he was very gentle with children, women and animals but had fits of anger and outrage if he felt betrayed. [3] His personal life was rather spartan, he was a vegetarian, did not smoke or drink alcohol, but he liked chocolates. During the war he developed a habit of working throughout the early hours of the night and then sleeping till late in the morning.

Much is made about his sex life, yet it seemed to have been quite normal. As a schoolboy he liked to look at pretty girls like other boys. In 1925 he fell in love with the seventeen-year-old daughter, Geli, of his half-sister Angela Raubal, who was then his housekeeper. He installed Geli in his apartment and she took singing and dancing lessons.

Then on September 18, 1931 she committed suicide while Hitler was traveling. He took it very hard. His later and final love affair was with Eva Braun, then an assistant to Heinrich Hoffmann, his photographer. He had a neat personal appearance but had bad teeth which he tried to hide when he laughed.

Despite having been only a corporal during the First World War, he seemed to have a good understanding of military tactics, at least where land armies where involved. After the war, the German Colonel-General Jodl unashamedly told his interrogators,

> Looking at the whole picture, I am convinced that he was a great military leader. Certainly no historian can say that Hannibal was a poor general just because ultimately Carthage was destroyed. [4]

As far as foreign policy was concerned, he quite openly pursued the recovery of German territory lost during World War One and the repudiation of the conditions imposed on Germany by the Treaty of Versailles. His first step was, on July 3, 1936, to reoccupy the then de-militarized Rhineland (the German area west of the Rhine and bordering on France). He had already established universal military service on March 16, 1935, and had started to build an air force. He accomplished this despite feeble protests from both England and France, the latter being beset by divisive internal political problems. Both Western Allies had neither sufficient forces nor public support for intervention. As a matter of fact, the French General Gamelin at that time did not posses one single unit ready for combat. [5]

Hitler's first attempt to try out his armed forces came in 1936 with the outbreak of the Spanish Civil War. Here Hitler sent in 1936 the first, thinly disguised "volunteers" to Spain. These were mostly air troops with warplanes. Italy under Mussolini also sent troops in support of General Franco. The opposition, the Socialist

Government, then in power, received substantial support from Russia, under the leadership of Stalin. This civil war lasted till March 28, 1939 and ended in a victory for the conservative forces under General Franco. This war proved to be a trial run for World War Two, which started only six months later.

In March 1938 Hitler overcame the (British Secret Service supported) resistance by the Austrian Government and annexed Austria, albeit with the overwhelming support of the Austrian population.

Hitler's next aim was to recover the Sudetenland (the former German border area that was given by the Allies to the then Czechoslovakia). He accomplished this with a treaty on September 29, 1938 between Premier Daladier of France, Mussolini of Italy and Chamberlain of Great Britain. By now the Western Allies were openly disturbed, not only by Hitler's flagrant violation of treaties, but also by the size of the German armed forces and the growth of Germany as an economic power. What is now decried as "appeasement" on the part of Chamberlain was in reality a policy of "stalling for time". The later Allies realized quite well that war with Hitler was unavoidable, but war could not be waged at that time (1938) due to insufficient strength on the part of the Allies.[6] It was now that both the British and the French Governments started massive re-armament programs and began to prepare their people for war. For example, there was no support for war in 1938 within the British Commonwealth countries, yet there was that support a year later, in 1939.[7]

Hitler's intervention in Czechoslovakia was not entirely unopposed within Germany. For example, General Ludwig Beck, the head of the German General Staff, resigned on August 27, 1938 in protest at the planned invasion.

This opportunity for armed intervention opened when Hitler, in 1939, voiced new demands, this time to reincorporate the former German city of Danzig into the Reich and further establish a land corridor between Germany proper and the province of East Prussia, then surrounded by Polish territory. The Polish Government strongly resisted, backed by a military treaty with England. Nevertheless, after making a nonagression treaty with Russia in August 1939, which included the Russian supply of raw material, which Germany lacked, Hitler ordered his armies to invade Poland on September 1, 1939. This now was the *casus belli* and both France and Britain declared war on Germany on September 3, 1939. World War Two had started.

It is now generally recognized that this war was in reality an extension of the First World War involving basically the same players.

While the German public was understandably quite disturbed about yet another war, the short duration (21 days) of the Polish campaign made it acceptable.

During 1940 Hitler secured his shipping routes for Swedish iron ore by occupying Denmark and Norway despite the valiant defense of Norway by British troops. This was followed by a lightening strike at France in May 1940. Hitler was acting throughout as supreme military commander planning and organizing each campaign to the last detail. So far, all his gambles had paid off, despite some strenuous objections by his generals. After the defeat of France, he tried to make peace with Britain which, at that time, was on its knees.[8] He even went so far as to let the British expeditionary forces escape from Dunkirk in France as a goodwill gesture. [9]

There followed a heated debate, regarding Hitler's peace offer, in the war cabinet where the proponents were lead by the Earl of Halifax. However, Winston Churchill, by then Prime Minister, prevailed in rejecting the peace feelers, relying instead on help from the United States. [10]

Hitler then considered an invasion of Britain, but noticed an unusually large concentration of Russian troops at the border of the then Russian occupied part of Poland, (in 1939 Poland was invaded by Russia, which thereafter occupied the eastern portion of Poland and then followed up with invasions of Finland and Rumania). Hitler surmised that Stalin was planning to invade Germany too[11] once Germany was occupied with war in the West (that such plans existed had been verified by former members of the Russian General Staff after the war). These observations were followed up by demands for new concessions on the part of the German Government, favoring the Russians. The Russian Foreign Minister handed these demands over during a visit to Berlin in 1940. This posed a dilemma for Hitler. He rejected the

Russian demands but decided to gamble again and planned an attack on the assembled Russian Army starting on May 15, 1941, rather than wait to be attacked himself. Unfortunately for him, Mussolini's reckless invasion of Albania and Greece forced him to rescue Italy first, causing him to delay the Russian campaign till June 15, 1941. This, in hindsight proved a fatal mistake since it did not allow sufficient time to conquer Moscow before the onset of the winter.

Nevertheless, he caught most of the Russian Army in the open near their Western borders without strong defensive positions (not needed if you plan to attack yourself). As a result, the German Army took over two million Russian prisoners during the first weeks of the campaign. However, the hard winter of 1941, Stalin's appeals to the patriotism of the Russian people, and his relocation of factories to the east of the Ural Mountains, finally turned the tide and at last stopped the German armies. The entry of the U.S. into the war following the Japanese attack on Pearl Harbor in December 1941 then brought badly needed war material into Russia and the Russian Army went on the offensive. The turning point came in the German defeat at Stalingrad on February 2, 1943, a bitter battle in which Germany lost 252,000 soldiers. Russia had recovered all of their lost territory by 1944 and by the end of that year even reached German territory. Russia received additional help from the U.S. and the British Armies, when they invaded France on June 6, 1944, forcing Hitler to divide his already decimated armies.

Other military setbacks, such as the defeat of the German North-Africa Corps at El-Alamain, by the British, led to a retreat from Africa and additional German losses.

The final German offensive in December 1944, commonly known as the "Battle of the Bulge", tried to stem the tide of American troops, now entering German territory from the West. This effort too collapsed and the war entered its final phase.

On April 11, 1945 the U.S. Army and the Russian Army met at the Elbe River and cut Germany in two, sealing Germany's fate.

The last defensive lines of the German Army at the River Oder were then breached by two Russian Armies on April 16, 1945 and the final Battle for Berlin began.

Notes

[1] Lucas, John, *The Duel*, Ticknor & Fields, Houghton Mifflin Company, 1991

[2] *Chronik Der Deutschen*, *Chronik Verlag*, Germany, 1983, p. 926

[3] Much has been made of his temper-tantrums, yet most of it was play-acting to drive home a point. Fest, Joachim C., *Hitler*, A Harvest Book, Harcourt Inc., 1973.

[4] Overy, Richard, *Interrogations, the Nazis in Allied Hands 1945*, Penguin Putnam, Inc., 2001

[5] Brendon, Piers, *The Dark Valley*, Alfred A. Knopf, New York, 2000

[6] Lucas, John, *The Duel*, Ticknor & Fields, Houghton Mifflin Company, 1991

[7] Lucas, John, *The Duel*, Ticknor & Fields, Houghton Mifflin Company, 1991

[8] Correlli Barnett, *The Collapse of British Power*, Sutton Publishing Ltd., 1997, p. 591.

[9] Lucas, John, *The Duel*, Ticknor & Fields, Houghton Mifflin Company, 1991

[10] Lucas, John, *The Duel*, Ticknor & Fields, Houghton Mifflin Company, 1991

[11] See also Irving, David, *Hitler's War*, Vol. 1, The Viking Press, New York, 1977, p. 205

APRIL 1945: LAST DAYS IN BERLIN

Let's start with Friday April 20, 1945. In order to follow the trail of the crime, as it were, we have to study what happened in the bunker during the last days in April. This bunker, located next to the Reich Chancellery, was originally constructed as an air raid shelter. It had two levels; Hitler and his closest staff used the lower level. The rear exit of the lower part of the bunker led into a large garden.

On April 20, 1945, Hitler celebrated his birthday in what was left of the Reich Chancellery. Here he was still surrounded by most of his closest associates even though the Russian Army was advancing rapidly on Berlin. Yet there were still roads open out of the city. The atmosphere at this party was understandably subdued. Still, champagne and canapés were served; Hitler gave a number of short speeches and, after the reception, reviewed a group of Hitler Youths that were decorated for bravery. German newsreel and press photographers duly recorded this event.[1]

The news from the front was bad. Two Russian armies had achieved a major breakthrough and were marching on to Berlin. Despite the bad news, Hitler was still very much in charge and seemed in full possession of his mental faculties. For example, during his daily military conferences he issued precise and detailed military orders to his commanders. On April 21, Hitler was trying to set up a counter-offensive by elevating General Steiner to the head of an army, and gave him the 4th SS police division, the 5th Jaeger division, and the 25th Panzer Grenadier Division.[2] This is quite in contrast to the "Hitler" we shall encounter after April 22.

Here is a list of the more important figures of his court that were present during Hitler's fifty-sixth birthday celebration on April 20, 1945 in the New Chancellery in Berlin and who would have been able to recognize any changes in Hitler's personality or appearance after April 22, 1945, the day of Hitler's escape from Berlin and of the subsequent entrance of his "double".

Marshall Hermann Goering; drove to Bavaria on April 20, 1945, arrived April 23, 1945. The head of the German Air Force; later committed suicide prior to his execution.

Armaments Minister, Albert Speer; left Berlin April 23, 1945 at 3 A.M. for Bad Wilsnack. He was later sentenced to 20 years in prison in Nuremberg.

Minister of Interior, Heinrich Himmler; departed April 22, 1945 for Hohenlychen, he was also the head of police and Security Services. He committed suicide on May 23, 1945.

Minister for Propaganda Josef Goebbels; committed suicide on 1st May1945.

Foreign Minister Joachim von Ribbentrop; Left Berlin on April 22, 1945; he was convicted and executed in October 1946.

Martin Bormann, Hitler's secretary; he was killed trying to escape Berlin on May 2, 1945.

Albert Bormann (brother of Martin Bormann); he flew out of Berlin on April 21, 1945.

Air Force Adjutant of Hitler, Colonel Nikilaus von Below; flew out of Berlin on April 29, 1945.

General E. Christian; left by plane for Bavaria on April 23, 1945.

Dr. T. Morell, Hitler's personal doctor; he departed April 22, 1945, for Berchtesgaden.

General Hans Baur, Hitler's chief pilot; attempted escape on May 1, 1945, he was wounded and taken prisoner.

Eva Braun, Hitler's mistress and later his wife; assumed to have left with Hitler on April 22, 1945.

Admiral K. von Puttkammer, liaison officer for the Navy; left April 22, 1945 for Berchtesgaden.

Field Marshall Wilhelm Keitel, head of the General Staff, ordered out of Berlin on April 22, 1945; he signed capitulation of Germany, was executed in 1946.

Colonel-General Alfred Jodel, chief of operations of the Armed

Forces; he was ordered out of Berlin on April 22, 1945, he was executed in 1946.

General H. Krebs; committed suicide on May 2, 1945.

SS General W. Burgdorf; assumed to have committed suicide on May 2, 1945.

General G. Christian; ordered out by plane on the night of April 23, 1945.

SS General Mohnke; captured on May 2, 1945.

SS Lieutenant General H. Fegelein, Hitler's liaison to the SS Army; he left the bunker on April 25, 1945, then disappeared.

Johanna Wolf;[3] left by plane on April 22, 1945 for Berchtesgaden.

Christa Schroeder;[4] left by plane on April 22, 1945 for Berchtesgaden.

Gerda Christian;[5] escaped Berlin to West Germany May 1, 1945.

Gertrud Junge;[6] escaped the bunker on May 1. 1945, remained in Berlin.

SS Major O. Guensche, adjutant of Hitler; captured May 2, 1945.

SS General Schaub, Hitler's chief adjutant; ordered out on April 22, 1945 to burn Hitler's papers.

Ambassador W. Hewel, Hitler's liaison to the foreign office; committed suicide on May 2, 1945.

Colonel Dr. W. Stumpfegger, personal physician of Hitler's double; committed suicide May 1, 1945.

Major General J. Rattenhuber, head of Hitler's security (*Begleitkommando)*; wounded and captured on May 2, 1945.

SS General H. Mueller, head of the GESTAPO (Secret Police); escaped on April 29, 1945 by plane to Switzerland;

Heinz Linge, Hitler's valet; captured on May 2, 1945.

All in all about 80 staff members left on the evening of April 22, 1945.

As can seen from the above list, only a handful of people remained in the bunker after April 22 – a date that is very important, as we shall learn later – who knew Hitler on a personal basis and who would recognize his voice or mannerism. This was important to avoid the discovery of a double in their midst. These were:

Dr. Goebbels, Bormann, General Krebs, Major Guensche, Linge (Hitler's valet), SS General Mueller (head of the GESTAPO) and Baur (the chief pilot).

In addition there were Rattenhuber, the Chief of Hitler's personal bodyguard, Walter Hewel from the foreign office, SS General Fegelein and the two remaining secretaries, Gertrud Junge and Gerda Christian.

The only senior officers present at the daily military briefings after April 22, 1945, other than General Krebs, were Generals Reimann, Weidling and Mohnke. Neither Weidling nor Reimann knew Hitler very well. These were local commanders, only concerned with the defense of Berlin.

We know that Dr. Goebbels, Hewel and General Krebs committed suicide and that Martin Bormann was killed while trying to escape. This leaves only Rattenhuber, Linge, Baur, Mohnke and the two secretaries as potential witnesses to the identity of the person supposed to be Hitler after April 22, 1945. However, Rattenhuber, Baur, Mohnke and Linge were very loyal followers and any secret would have been safe with them. As to the two secretaries, General Mueller told his interrogator[7] "that one of them was older and very loyal while the other one was very young and stupid." Hermann Fegelein, Eva Braun's brother-in-law, was a special story. We learn more about him later.

The days of April 21 and 22 saw all kinds of unusual activities within the bunker while the first Russian artillery shells fell on Berlin. This is especially true of Sunday, April 22, 1945, a key date.

At the noon military conference, Hitler became quite upset and finally ordered all to leave, except for Keitel, Jodel, Krebs, and Bormann.[8] According to later reports, Hitler, at that private meeting, relinquished supreme command of the armed forces and ordered both Keitel and Jodl to leave Berlin immediately for Bavaria and to assume command of what was left of the German forces. The two generals then left Berlin on that day – April 22 – but decided to stay in northern Germany.[9]

This was a momentous decision for Hitler to relinquish command at that time. Why did he not wait? However, it does make sense if he was planning to leave Berlin on that very same day.

Hitler then ordered Schaub, who had a key to Hitler's safe, to burn all papers that were in the safe located in the bunker and then fly to Berchtesgaden to destroy all files at Hitler's mountain retreat. This was done the next day. Hitler also gave Schaub several 100,000 German marks in cash for distribution to Hitler's and Eva Braun's relatives.

In the afternoon he had tea with Eva Braun and two of his secretaries. After tea he bid farewell to his secretaries and ordered them to fly out of Berlin the same evening.[10] He said the situation looked hopeless. Nevertheless, the girls pleaded with him that they wanted to stay. To this Hitler replied: "Ah, if only my generals were as brave as my women." He then stood up and kissed Eva Braun on the lips. One of his secretaries, Ms. Christian stated that this was the first time she ever saw Hitler do this.

At 5 P.M. he then called Goebbels by telephone, who arrived at about 6 P.M. at the bunker, together with his family and a mysterious female. Albert Speer, his Armament Minister, and his Foreign Minister, Ribbentrop, also came and said their goodbyes. However, Speer decided to stay till early next morning, departing only around 3 A.M. on April 23, the day when everything in the bunker changed.

Notes

[1] American Heritage, *Pictorial History of World War II*, Heritage Publishing Co., Inc., 1966, p. 574

[2] Ziemke, Earl F., *Stalingrad to Berlin, the German Defeat in the East*, Center of Military History, U.S. Army, Washington, DC, 1968, p. 477

[3] One of Hitler's secretaries.

[4] One of Hitler's secretaries.

[5] One of Hitler's secretaries.

[6] One of Hitler's secretaries.

[7] Douglas, G regory, *Gestapo Chief, The 1948 Interrogation of Heinrich Mueller*, James Bender Publishing, 1995

[8] O'Donnell, James, *The Bunker*, Da Capo Press, 1978

[9] *Chronik Der Deutschen*, *Chronik Verlag*, Germany, 1983, p. 926

[10] Christa Schroeder, *Er War Mein Chef,* second edition, Georg Mueller Verlag, Germany, 1985

THE ESCAPE

As described in the book *Gestapo Chief*[1], it all started on a cold day in March 1945 when Hitler summoned SS General Heinrich Mueller, the head of the feared GESTAPO, for a private talk in the garden of the Chancellery during one of Hitler's customary half hour walks. The day was cold and windy. There were guards but they stayed away, besides Hitler spoke in a low voice to make sure nobody would hear. After a discussion of the military situation, which even Hitler realized as being hopeless, Hitler then asked Mueller about how it all should end and he listed as options:

A. Surrender;
B. Go to the mountains and keep on fighting; or,
C. Commit suicide.

Mueller strongly suggested to Hitler not to surrender, there was no point to it. Going to the mountains would only prolong the inevitable. This left suicide, but here Hitler had trouble accepting this idea.

He finally asked Mueller what he would do in his place. The answer was: try to escape.

This seemed agreeable to Hitler and he mentioned Switzerland as a likely place to go, but Mueller talked him out of it fearing that the Swiss might extradite him to the Allies. He suggested Spain instead, and specifically Barcelona. This city had a port and it would be easy later to smuggle him out on a ship to South America with the help of Mueller's local agents.

This plan then was discussed in detail and culminated in the suggestion to have Hitler flown out of Berlin to the South of Germany and from there to Barcelona, using Werner Baumbach, a loyal and much decorated German air force pilot. It certainly would have looked suspicious if Hitler used Hans Bauer his chief pilot on this occasion (this is exactly what the Russians suspected of him afterwards, see chapter on interrogations).

Mueller then indicated that in order for the plan to succeed it was of the utmost importance to have proof that Hitler was dead. Here Hitler mentioned the use of his double – causing both men to laugh.

Mueller's proposal was to send all the staff i.e. secretaries, doctors, clerks etc. out of Berlin just after the April 20, celebration of Hitler's birthday.

All the other important people like Goering and Ribbentrop would be leaving too. There would remain only relatively unimportant people in the bunker besides Goebbels, who already indicated he intended to commit suicide. If the Russians found the corpse of Goebbels, that would lend an air of credibility to the whole affair.

Returning now to the evening of April 22, Hitler said goodbye to two of his secretaries, his adjutant Albrecht, Admiral von Puttkammer and Albert Bormann. Between 9 and 10 P.M. nine of the ten planes with all the staff and baggage had left Berlin and later arrived safely in Salzburg in Austria. However, the tenth plane with Hitler's important papers and one of his valets on board crashed en route and burned. This also destroyed part of Eva Braun's jewelry.

At about 8.30 P.M., SS General Mueller observed Hitler accompanied by his favorite shepherd dog, Blondi, leaving the bunker and walking towards the garden. Mueller then met and talked to him there. Also present was Linge, Hitler's valet, and Rattenhuber, his security chief. There was nothing unusual about Hitler walking his dog that evening. Security personnel as always restrained the guard dogs and switched off the floodlights. It should be noted that only the lower bunker floor, where Hitler lived, had an emergency exit to the garden. Thus few people would have been able to notice his customary walks outside. Hitler now thanked Mueller warmly for all he had done for him and for Germany and he asked Linge to give Mueller a leather briefcase containing a large sum of Swiss francs, a personal letter and a medal. Then Hitler departed through the rear exit of the garden. Linge soon returned, he was crying and said, "The chief is gone. I will never see him again."

Rattenhuber too reappeared and said to Mueller: "The Chief is gone but now we have a new Chief," meaning Hitler's double. According to Mueller, the second "Hitler" then arrived with a similar shepherd dog taken from the kennel (there were two dogs left) and entered the Bunker accompanied by Rattenhuber. Prior to entering the bunker on April 22, the double was kept at the hotel Kaiserhof[2] in Berlin near Mueller's office.

The question now is: Where did Hitler go that evening?

According to Mueller (who after all should know best, having been the organizer of the whole affair) Hitler left the garden by the rear door and flew out of Berlin aboard a Type Fa 223 twin-rotor helicopter for Hoerching airfield near Linz, Austria.[3] Mueller stated that U.S. troops later discovered this machine there. This might not be true. A type Fa 223 V51 (Serial Number 233 000) helicopter piloted by Otto Dumke arrived around that time in Ainring near Salzburg supposedly coming from Rechlin the big German airbase near Berlin. The U.S. Army later captured this machine in Ainring. Ainring then was the home base of the helicopter Transportation Squadron 40, which owned several type Fa 233 copters.[4] It seems quite possible that this aircraft returned to its home base at Ainring after unloading Hitler in Linz-Hoerching (about 70 miles away). This particular copter had a load capacity of 1320 lbs, a cruising speed of 87 mph and a range of 272 miles (without auxiliary fuel tanks). It should be noted that Germany had helicopters since before the war.[5] As a matter of fact, the famous aviatrix Hanna Reitsch flew one *inside* the Sportspalast in Berlin in 1938. She was also planning to fly by helicopter to Berlin on April 26, together with General Ritter von Greim but found that the requested copter was unavailable. She used a spotter plane instead.

From Hoerching, Hitler and at least two other passengers departed in a four engine Ju 290A airplane for Barcelona, on April 26, 1945 at about 8 P.M. It landed there on April 27, 1945. Mueller stated to his U.S. interrogators that he received confirmation that the plane landed safely.[6]

Incidentally, there are photos in the book, *Gestapo Chief*, of both the helicopter and the airplane. The latter is shown with Spanish markings (the German ambassador gave the plane to the

Spanish government). In his book Gregory Douglas claimed that Hitler escaped in a Junkers 290 A6 plane, Serial number 0185. Since Mueller did not give this information to his interrogators, we must assume that this identification was based on Douglas' own research.[7] However, this conflicts with information given by Sweeting in his book *Hitler's Squadrons*, who claimed that this particular aircraft was badly damaged beyond repair at the Russian front in May 1944.[8]

However he stated that there was another JU 290 A-3, Serial number 0163, Code PI PQ, that was later located at the seaport of Travemunde[9] near Hamburg. These planes, with special fuel tanks, could have a range of up to 4225 miles. What is intriguing is, that this is exactly the airport on which Colonel Baumbach landed on April 28, 1945 before he established contact with Admiral Doenitz, Hitler's successor.[10] It is quite likely that Douglas was misinformed and that Colonel Baumbach, who, at the beginning of April, was made commander of Hitler's Flight Command,[11] flew this plane (Ser. No: 0163) with Hitler from Linz-Hoerching to Barcelona on April 26, 1945 as Mueller claimed.[12] This plane then landed in Spain on April 27, and after deplaning Hitler and his followers, was flown back to Germany and more specifically, to Travemuende on April 28. The plane later was blown up at this airport on May 3, 1945 prior to the arrival of British troops.[13]

Such an explanation makes sense, since it fills neatly in the time frame between April 21 (the day prior to Hitler's escape from Berlin) and April 28, when Baumbach claimed he flew into Travemuende. The explanation of what happened during these days was left out in Baumbach's memoirs. The reader may guess the reason why.

The aircraft shown in the Douglas book, having Spanish markings, apparently was another A290 plane which arrived on April 5, 1945 on a scheduled flight from Germany and operated by the German airline Lufthansa.[14] Incidentally, this was the last commercial flight from Germany to Spain.[15]

What is less clear was, who accompanied Hitler. The original flight order listed, besides Hitler and Eva Braun, Dr. and Mrs. Goebbels, Bormann, General Mueller, General H. Fegelein, Dr.

Stumpfegger, Ambassador Hewel, Colonel Betz and General Burgdorf. We now know that at least five of the last eight listed above did not go. However, Mueller insisted that Eva Braun and SS General Hermann Fegelein were on the plane. One likely passenger who was not on the original flight order was SS General Dr. Hans Kammler (sometimes called "Hitler's gray eminence") who was practically unknown to the public. Kammler, at that time, was head of all advanced weapons research ranging from four-stage rockets to long range artillery and, most importantly, all aspects of the German effort to build an atomic bomb. General Kammler disappeared without a trace after April 23, 1945 from Prague, which is close to Hoerching airfield. This may have been the result of a long and private meeting with Hitler on April 3, 1945. Incidentally, in his book, Philip Hentschel also mentioned the landing of the Ju 290 airplane in Barcelona on April 27, 1945, although he had no idea who the passengers were.[16]

Even more of a mystery is the question of what happened after the plane landed. It is quite feasible that General Franco, the Spanish dictator, did hide Hitler and his entourage as a "thank you" for helping him win the Spanish Civil War between 1936 and 1939. It should be noted that Franco could not have won the civil war without Hitler's military support. This should have made General Franco very grateful indeed. Besides, Franco was an honorable man. There is also a strong possibility that Hitler could have gone by ship to Argentina from Barcelona, a possibility strongly suspected by Stalin. One has to understand that there existed a very efficient organization in Spain and later in Argentina that, according to Uki Goni ferreted hundreds of German, Austrian, French, Belgian, Dutch, Slovakian and Croatian war criminals and collaborators to Argentina between 1945 and 1950.[17] A good portion of these criminals had been condemned to death in their home countries. This rescue organization was originally started by SS Leader Heinrich Himmler's special envoy Carlos Fuldner, who arrived in Madrid, Spain, on March 10, 1945 with plenty of funds. This effort was further supported by the then Argentine Head of State, Juan Perón, and by prominent leaders of the Catholic Church. These refugees arrived in Buenos Aires either by airplane from Madrid or by ship from Spain. Additional venues of escape opened up, after 1947, through Italy and Switzerland. With the exception of only two of the war criminals (Eichmann and Priebke) all enjoyed a life of peace and comfort, mostly under false identities.

It is therefore quite possible that Hitler and his wife could have submerged themselves into such a sub-culture.

The original flight order listed Colonel Baumbach as the pilot of the plane to Barcelona. Werner Baumbach was a highly decorated German bomber pilot during the war and during the last months of the war, was the commanding officer of Hitler's personal fleet of airplanes. It is very significant, that on April 21, 1945, the day before Hitler left Berlin, Baumbach requested and received a certificate from the German Air Ministry, signed by a Colonel Wiltner, stating that Baumbach was qualified to work as a civilian pilot and that he could call himself a "Flight Captain".[18] This was no doubt in preparation for piloting civilian airplanes in Spain or South America. As a matter of fact, Baumbach later had a fatal airplane accident in 1953 in Argentina. This brings up the interesting question: did Hitler too, travel to Argentina from Barcelona?

In his book *Broken Swastika*,[19] Baumbach mentioned that on April 21, 1945 he was in Berlin where he received a letter from Speer. Curiously, the next paragraph was completely out of context and stated: "We had fixed up long-range aircraft and flying boats that could take us anywhere on earth". He then switched the subject to say that he was in northern Germany on April 28 and 29, arriving at the Travemuende airport (near Hamburg). There is no explanation of how he left Berlin, or what he did between April 21 and April 28. It seems quite possible that he flew back to Germany from Barcelona on the 27 or 28 of April, after having left the Hitler party in Spain.

It should be noted here that Hitler's plane was not the only one going to Spain. On May 8, 1945, Albert Speer's private "Condor" plane flew from Oslo, Norway to near San Sebastian, Spain, with the escaping Colonel Leon Degrelle, the leader of the Belgian SS volunteers, who fought the Russians[20] during the Second World War.

The next day after Hitler's departure, Monday, April 23, saw many changes. Most of the staff was gone, there was less security, discipline was relaxed and, most of all, instead of an energetic "Fuehrer", there was a feeble, sickly and ignorant substitute who had to be kept isolated and who had to be coached on what to say by Goebbels and Bormann.

Here is a statement from Captain Helmut Beermann of the Security Detail that gives some insights of what the situation was like following Hitler's departure:

> We veterans of the bunker called this day Blue Monday because, with the departure of half of our comrades and the arrival of the whole Goebbels family and Eva Braun, everyone could now read the writing on the wall. The last act was about to begin. My own dream of again seeing Berchtesgaden vanished. Colonel Schaedle, my commanding officer, insisted I stay, since I was by now an old hand in the outfit. Previously, I had enough officers and men to be able to work out shifts, twelve hours on, twelve hours off. Now, every person was assigned to his task for the duration, which might be for two days or for two weeks. I issued sleeping bags, so that some of my key men could sleep at or near their stations. The old spit-and-polish discipline was all shot to hell. Many soldiers were not even saluting anymore.[21]

There are two interesting items in this statement, first the arrival of Eva Braun.

This must have been a "new" one (perhaps a double), since the real Eva Braun had lived within the bunker already since April 15, 1945, after leaving her apartment at the Chancellery. Secondly, we note here the beginning of a breakdown of morale and discipline. Did the soldiers know about Hitler's departure? Even if they knew no specifics, then rumors certainly must have abounded.

Notes

[1] Douglas, Gregory, *Gestapo Chief, The 1948 Interrogation of Heinrich Mueller*, James Bender Publishing, 1995

[2] This probably was the origin of Chief Pilot Bauer's later statement to the Russians about the rumor of a "porter" at the Kaiserhof having great similarity to Hitler.

[3] Hoerching is now the municipal airport of the city of Linz in Austria.

[4] Coates, Steve, *Helicopters Of The Third Reich*, Ian Allan Publishing Ltd.

[5] Winters, Jeffry, *Served Straight Up,* Supplement of *Mechanical Engineering Magazine* (100 years of flight), ASME, December 2003, p. 20

[6] Douglas, Gregory, *Gestapo Chief, The 1948 Interrogation of Heinrich Mueller*, James Bender Publishing, 1995

[7] Douglas, Gregory, *Gestapo Chief, The 1948 Interrogation of Heinrich Mueller*, James Bender Publishing, 1995

[8] Sweeting, C. G., *Hitler's Squadron, The Fuehrer's Personal Aircraft and Transport Unit, 1933–1945*, Brassey's Inc., 2001

[9] Werner Baumbach, *Broken Swastika*, Dorset Press, 1992

[10] Baumbach, Werner, *Broken Swastika*, Dorset Press, 1992, p. 193

[11] Sweeting, C. G., *Hitler's Squadron, The Fuehrer's Personal Aircraft and Transport Unit, 1933–1945*, Brassey's Inc., 2001

[12] Douglas, Gregory, *Gestapo Chief, The 1948 Interrogation of Heinrich Mueller*, James Bender Publishing, 1995

[13] Sweeting, C. G., *Hitler's Squadron, The Fuehrer's Personal Aircraft and Transport Unit, 1933–1945*, Brassey's Inc., 2001

[14] Lufthansa Airlines operated three type Ju 290 aircraft.

[15] Sweeting, C. G., *Hitler's Squadron*, The Fuehrer's Personal Aircraft and Transport Unit, 1933–1945, Brassey's Inc., 2001

[16] Henshall, Philip, *The Nuclear Axis, Germany, Japan and the Atomic Bomb Race*, Sutton Publishing Limited, 2000

[17] Goni, Uki, *The Real Odessa*, Granta Books, London, 2002

[18] HERMANN HISTORICA, 45th Auction Catalog for October 17–18, 2003, pp. 346–341, "Oberst Baumbach Memorabilia"

[19] Baumbach, Werner, *Broken Swastika*, Dorset Press, 1992

[20] Toland, John, *The Last Days*, A Bantam Book/Random House, Inc., 1967

[21] Joachimsthaler, Anton, *The Last Days Of Hitler*, Cassell & Co., London, 1995

A NEW "HITLER"

Let's now discuss the person left in the bunker on the evening of April 22, to pretend to be the "Fuehrer", in order to cover up Hitler's escape and to play his part in what truly was an amazing charade. I am talking about Hitler's double.

The only information I have found on this subject is from SS General Mueller[1] and this may not be exact. However, we know from the Russian autopsy of the body of the "double" and from the Nordon Report[2] that the major parts of Mueller's description can be confirmed. Finally, there is the well-publicized photo of the "double" taken by the Russians on May 2, 1945 after his corpse was discovered buried in the bunker garden.

SS General Mueller stated[3] that he found this man in 1941 and that he was born to the Sillip family in the Waldviertel district of Austria. He was a distant relative of Hitler and worked in Breslau, was a party member, a member of the SA and was unmarried. While he was not "brilliant", he was quite easy to work with. The resemblance to Hitler was remarkable but he was too short. According to an unconfirmed statement attributed to Eva Braun, he was 5 cm (2") shorter (see also the Russian findings, as shown in the Nordon report).[4] The next problem was his ears. These did not match Hitler's but there was nothing that could be done about this. He also smoked. When he stopped smoking, he gained weight so he had to be put on a diet. The next problem was his accent, which had to be resolved, together with the problem of his height. Here special soles were put on his boots. He learned to use Hitler's favorite phrases when he talked and was taught his mannerisms. For example, when Hitler laughed, he had a habit of holding his hand across his mouth so his bad teeth would not show. The double had met Hitler twice in order to study him close up. Hitler later remarked that when he saw the double, he felt as if he was looking into a mirror. Joachimsthaler tried to show proof that there was no double by quoting one of Hitler's secretaries, Johanna Wolf, that "…the Fuehrer would never have tolerated this."[5] This statement is probably true when viewed in the right context. Hitler certainly would have rejected a double being *in his presence*, but not when he would be at a different location. However, in the same book[6] we read about a statement by former SS guard Hans Hofbeck indicating that Chief Pilot Baur told him in a Russian prison, "that a man from Breslau had been presented who looked very much like Hitler."

According to Mueller, this double was used only a few times before and this was after the July 20, 1944, assassination attempt on Hitler's life.

Mueller further mentioned that Linge and Rattenhuber knew about the double as did Goebbels, but Bormann and Guensche were not told. It appears also that General Mohnke was somehow involved in the conspiracy due to his later cover-up actions.

Following his entry into the bunker on the evening of April 22, 1945, the new "Hitler" was kept very isolated and was allowed to see only few and select persons. Bormann, since he was not told, was somewhat confused by the double. He remarked to Mueller on April 23: "The chief looks very different, Mueller, do you think he might have had some kind of a stroke?" General Weidling, who did not really know the real Hitler, remarked to a Russian War correspondent, Lew Slavin, that he saw "Hitler" for the first time on April 24 1945, when he (Weidling) was appointed commander of the Berlin forces:

> As I now saw the Fuehrer I was astonished at his appearance, he was a human wreck, the head shook, his hands trembled and his speech was hardly distinguishable.

He added that,

> …there was an atmosphere of mistrust and there were likewise rumors abounding that proclaimed Hitler was no longer Hitler but his double!

To quote from Fest's book: "Since Hitler's return to the Chancellery, there was evidence that something secret was going on."[7]

Dr. Schenck who met the new Hitler on April 29 for the first time, stated, "The pathetic man that I saw bore little resemblance to the old, mesmerizing idol of the masses that was so familiar to millions."

At midnight on April 22, Albert Speer tried to talk to Hitler again before he returned to Wilnack, but he was told Hitler was asleep. This was quite unusual since the real Hitler hardly went to bed before 3 A.M. Another curiosity was that contrary to the real Hitler, his double had breakfast at 8.30 A.M. instead of around noon!

Also contrary to the real Hitler who gave very precise military orders to his staff and his generals, his double was not trained in the art of warfare. Therefore he only issued platitudes such as "you must quickly execute all relief attacks", or "advance on all fronts". Even Minister Goebbels' coaching did not help here having had no military experience himself, although Goebbels had many closed door conversations with this Hitler, according to Gerda Christian, one of the remaining secretaries.

As told by McKale, again, General Weidling described the new Hitler as "a sick weakling hardly able to stand up or walk, manipulated by Goebbels."[8] Others stated that "Hitler" often closeted himself with Bormann and Goebbels.

As the German magazine *Spiegel* reported on January 10, 1966 in an understatement: "His (Hitler's) span of practical concern has narrowed." Thus he did not impress the remaining local military commanders during the brief military situation conferences *after* April 22. To quote General Mohnke from April 29:

> Hitler's midnight briefing had been short, desultory, uninformative.

He further stated to O'Donnell:

> Throughout the whole talk I was sitting almost at his (the double's) side, perhaps three or four feet away. But he was either gazing at the wall or looking down at the floor. After the first ten minutes, our talk ceased to be a conversation.[9]

This description certainly does not fit the Hitler who was at the same place three days earlier!

On April 30, the double issued an order (undoubtedly dictated by Goebbels) to General Weidling to authorize a break-out of Berlin for the remaining troops. Curiously, this order was typed on Hitler's personal letterhead! An earlier breakout request, on April 28, by General Weidling, was brusquely rebuffed by Goebbels without even bothering to consult Hitler. This he would never have done had the real Hitler been present!

Here is an apt description by Dr. Schenck who, on April 29, was asked by Prof. Haase to consult with the double, taking two nurses along. He testified:

> I was shaken a bit since I had never seen my Fuehrer before, except from an admiring distance. I knew, of course that this was Adolf Hitler and no Doppelgaenger...

(this was rather disingenuous, he must have said this to please his interrogators, since he had no way of knowing the truth). He then continued:

> Hatless, he was still wearing the familiar, once spotless, natty gray tunic with green shirt and long black trousers. He wore his golden party badge and his World War One Iron Cross on his left breast pocket. But the human being buried in these sloppy, food-stained clothes had completely withdrawn into himself. I could see his hunched spine, the curved shoulders that seemed to twitch and suddenly to tremble. He struck me as agonizing. Hitler seemed hardly able to shuffle the two paces forward to greet us. His eyes, although he was looking directly at me, said nothing. They did not seem to be focusing. The whites were bloodshot. His handshake was listless. At fifty-six the Fuehrer was a palsied, physical wreck, his face wrinkled like a mask, all yellow and gray.

General Mueller hinted to his interrogators, "the double was drugged, likely by Prof. Haase."[10] This may explain the new Hitler's obviously listless impression he made on Schenck and others. Drugging was also suspected by Ambassador Hewel, who was reported to have said that Professor Haase may have given

Hitler strong tranquilizing shots since the Fuehrer had very calm, even placid, periods during the last days.

The description of this Hitler's soiled uniform is rather revealing. This is in stark contrast to the real Hitler who was very particular when it came to his personal appearances. He also was very keen on personal hygiene. For example, he would always rinse his mouth after every meal.[11]

Schenck stayed in the bunker and later that night, after some drinking bouts, had to visit a toilet. The upper toilet happened to be clogged so he proceeded to walk to the lower level of the bunker towards the Fuehrer's quarters. To his astonishment, the two security guards normally guarding the entrance door were gone and he was able to walk by Hitler's sitting room unmolested, where he saw, what he believed to be, Hitler in an intense conversation with Prof. Haase.[12] It is interesting to note here how security had lapsed since the departure of the real Hitler!

Dr. Schenck observed, that while sitting on a table, the new Hitler held his steel rimmed glasses in his hand. This is another curiosity since the real Hitler's glasses were nickel rimmed. As Minister Speer noticed, there was a distinct change in Hitler's daily routine. Now there were no more of the usual midnight conferences that Hitler was so fond of. Once the busiest of hours, now at night, there were few souls about in the lower bunker.

During his walk-by observation, Dr. Schenck diagnosed that the substitute Hitler suffered from Parkinson's disease! General Mohnke who knew the real Hitler quite well, later ridiculed that statement.

Let's compare the above physical condition and appearance of the Hitler double to those of the real Fuehrer:

> The real Hitler did have some slight tremors in his left arm and left leg ever since the bomb explosion on July 20, 1944. He also suffered a sinus infection in September of 1944 followed by a bout of jaundice, which left him very weak and frail. However, he seemed to have recovered sufficiently so that he rarely saw a doctor during the first months of 1945. The only complaint was redness of the eyes and he had his valet Linge administer drops of cocaine solution.[13]

According to D. Irving describing the real Hitler:

> On March 30, 1945, after issuing a clear-sighted appraisal of the situation... His malevolently brilliant mind was still functioning.[14]

Similar statements were made by a Navy Commander Luedde-Neurath who attended the April 20 conference (when the real Hitler was still in Berlin) and who recorded in his diary: "Hitler's speech and eyes were as expressive as ever. His spiritual elasticity appeared preserved. He was not insane."[15] His doctors too were unanimous in agreeing that his sanity remained intact until the end, even though his blood-shot eyes (probably caused by the dusty bunker atmosphere) became so poor that he had to put on his spectacles even to read documents typed on special big-faced typewriters. As Fest writes:

> Frail as he was, he still preserved something of his magnetic powers...,

and he gives as an example:

> When in March of 1945 Gauleiter Forster visited Hitler and was very much in despair about the Russian attack on his city of Danzig. Yet after only a brief conversation with Hitler, Forster's mood was completely transformed towards the positive![16]

Notes taken by C. Schroeder, one of Hitler's secretaries, mentioned that her boss took anti-gas tablets and his doctors, Professors Brandt and Hasselbach told Hitler that this might have caused the occasional tremors in his left hand.

While it is possible for a double to fake a trembling hand, it is of course not possible to duplicate a brilliant mind. This was obvious from the vague and listless orders that the double of Hitler, issued. As a result, the generals sensing a power vacuum disobeyed all orders after the Fuehrer's April 22 departure and followed their own strategy of survival.

As to the physical impairments, we have photographs of the real Hitler inspecting a group of Hitler Youths receiving medals for fighting the Russian troops. Here he is seen standing, walking and touching some of the boys in his heavy overcoat on his fifty-sixth birthday on April 20, 1945.[17] Probably the last photo shown of Hitler pictures him standing erect and with his hands behind his back together with his adjutant, Schaub, while inspecting bomb damage at the Chancellery on April 21, 1945.[18] He also had no problems climbing the 48 steps of the heavy iron spiral staircase out of the bunker.

As to Hitler's stamina, I quote from J. Goebbels diary,[19] it states for March 30, 1945:

> I have the impression the Fuehrer is greatly overworked in the last few days. During the last 24 hours, for instance, he had only two hours of sleep.

And on April 3:

> He is doing the utmost to pull his military staff together and to fill them with confidence for the future. He is tirelessly preaching the spirit of battle and resistance…

It seems obvious that the new "Hitler" (the double) did not possess this energy, nor had he real power. An example of this was, that when Goering sensed the power vacuum in Berlin, he sent his famous telegram to the Fuehrer on April 23, giving what he thought was an ultimatum to Hitler to reply before 10 P.M., otherwise he, Goering, would take over as Chancellor of Germany. Bormann, who disliked Goering, immediately took this opportunity and (falsely claiming: in Hitler's name) ordered Goering to be arrested by the SS troops. Bormann's handwritten telegram survived the war according to D. Irving.[20] During these days (after April 22) Goebbels and Bormann were the real powerbrokers and made all the decisions. This is apparent from Bormann's signed message (not "Hitler's") to Admiral Doenitz during the morning of April 30 (prior to the supposed "suicide" of Hitler), to inform him that Himmler tried to make peace, using Sweden as mediator, against orders.[21] There seems also to be some questions regarding the famous testament of Hitler dated April 29, 1945[22] and witnessed by Bormann, Goebbels, Krebs and

Burgdorf.[23] According to G. Junge (Hitler's youngest secretary) Hitler, the double, dictated his testament to me from "notes". James P. Donnell strongly suggests that these were notes written by Goebbels, matching exactly his style and phraseology. As a matter of fact, Goebbels even went so far as to add his own political testament as an appendix to the "Hitler" document. There seemed to be a touch of vanity in this undertaking. He probably needed to explain why Doenitz was chosen as Hitler's successor instead of him.

Yet, when we look at Hitler's signature on the political testament,[24] we see the same small but, clearly written signature, so familiar from other previous documents. This could not have been signed on April 29, at four a clock in the morning by a tottering, unsteady and trembling "Hitler" as Dr. Schenck described him on the same day unless, of course, a facsimile signature was used![25] It is also unlikely that the substitute Fuehrer was up at four o'clock in the morning.

Here are the final bunker communications:

Martin Bormann, instead of Hitler, sent this telegram to Admiral Doenitz on April 30, 1945:

> The Fuehrer appointed you, Herr Admiral, as his successor in place of Reichsmarshall Goering. Confirmation in writing follows. You are hereby authorized to take any measures which the situation demands. Bormann.

The end came on April 30, in the afternoon. Here is the most likely scenario of what happened: Dr. Stumpfegger gave the double his daily customary injection, except this time it was most likely poison and the double soon slumped over on the sofa of the living room, dead. To make sure he was dead, General Rattenhuber probably then shot the double in the forehead, using the double's own 7.65 caliber Walter pistol.[26]

Now the double of Hitler was dead. It appears from Russian investigations, that he was poisoned and then shot in the forehead, to make sure. This confirms Rattenhuber's statement to his Russian captors that Hitler was shot. It also verifies the story told to his Russians interrogators by the SS Guard Ackermann.

On May 1, 1945, at 2.45 P.M., the following additional message was sent to Doenitz:

> Fuehrer died yesterday at 3:30 P.M. In his will dated April 29, he appoints you as President of the Reich, Goebbels as Reich Chancellor, Bormann as Party Minister, Seyss-Inquart as Foreign Minister. The will, by order of the Fuehrer, is being sent to you and to Field Marshall Schoerner and out of Berlin for safe custody. Bormann will try to reach you today to explain the situation. Form and timing of the announcement to the Armed Forces and the public is left to your discretion. Acknowledge. Goebbels – Bormann.

What is very curious about the latter message is, why would Hitler appoint Goebbels as Chancellor knowing quite well that Goebbels is going to commit suicide on the very same day!

Also, there was no hint of how Hitler died. Of course, the official suicide story only surfaced five months later when the apparent escape of Hitler became an embarrassment for the Allies.

Here are summaries of the salient behavior and appearances of both Hitler and his double from the observations of the surviving witnesses both shortly before and after April 22, 1945.[27]

HITLER BEFORE AND ON APRIL 22, 1945	HITLER AFTER APRIL 22, 1945.
Alert	Dull, drowsy
Could climb forty-eight steps to the bunker exit.	Hardly able to climb two steps
Could issue detailed military orders	Could utter only generalities
Slept late	Had breakfast at 8.30 A.M.
Preferred late night conferences	Had only short afternoon conferences
Issued verbal and written orders	Orders were issued only by Goebbels or Bormann in "Hitler's" name.
Had lunch with Eva Braun	Avoided all social contact with Eva.

Dresses neatly	Wore uniform soiled with food stains[28]
Took customary one-half hour walks in the bunker garden	Never left the bunker.

It seems impossible that such a drastic personality change could occur within less than twenty-four hours. We have therefore to assume that the Hitler that was present in the bunker between April 22 and April 30, 1945 was indeed a double.

It is puzzling indeed why the presence of the double of Hitler has gotten so scant attention by established historians despite the official Russian photographs of his existence.

Why, for example, were the witnesses never asked: "Who shot the double and who buried him?" Maybe the answer to all of this is that the whole story of Hitler's suicide may have collapsed if one were to admit that there was indeed a double in the bunker.

Notes

[1] Douglas, Gregory, *Gestapo Chief, The 1948 Interrogation of Heinrich Mueller*, James Bender Publishing, 1995

[2] Brown, Anthony Cave, *The Last Hero, Wild Bill Donovan*, Vintage Books, a division of Random House, 1984

[3] Douglas, Gregory, *Gestapo Chief, The 1948 Interrogation of Heinrich Mueller*, James Bender Publishing, 1995

[4] Brown, Anthony Cave, *The Last Hero, Wild Bill Donovan*, Vintage Books, a division of Random House, 1984

[5] Joachimsthaler, Anton, *The Last Days Of Hitler*, Cassell & Co., London, 1995, p. 181

[6] Joachimsthaler, Anton, *The Last Days Of Hitler*, Cassell & Co., London, 1995, p. 257

[7] Fest, Joachim C., *Hitler*, A Harvest Book. Harcourt, Inc., 1973

[8] McKale, Donald M., *Hitler The Survival Myth*, Cooper Square Press, 1981

[9] O'Donnell, James, *The Bunker*, DA CAPO Press, 1978

[10] Douglas, Gregory, *Gestapo Chief, The 1948 Interrogation of Heinrich Mueller*, James Bender Publishing, 1995

[11] Schroeder, Christa, *Er War Mein Chef*, second edition, Georg Mueller Verlag, Germany, 1985

[12] Prof. Haase was one of the substitute Doctors responsible for Hitler's health. In April 1945 he was in charge of the field hospital located in the cellars of the

Chancellery. Haase was taken prisoner by the Russian troops on May 3rd, was taken to Russia where he died in the Fall of 1945.

[13] Irving, David, *Hitler's War*, Avon Books, a division of Hearst Corp., 1990

[14] Irving, David, *Hitler's War*, Avon Books, a division of Hearst Corp., 1990

[15] McKale, Donald M., *Hitler The Survival Myth*, Cooper Square Press, 1981

[16] Fest, Joachim C., *Hitler*, A Harvest Book. Harcourt, Inc., 1973

[17] Taylor, Blayne, *Guarding The Fuehrer*, Pictorial Histories Publishing Company, 1993, pp. 242, American Heritage, *Pictorial History Of World War II*, Heritage Publishing Co. Inc., 1966, p. 574

[18] *Chronik Der Deutschen*, Chronik Verlag, Germany, 1983, p. 926

[19] Trevor-Roper, Hugh, *Final Entries 1945 The Diaries Of Joseph Goebbels*, G. P. Putnam's Sons, 1978

[20] Irving, David, *Hitler's War*, Avon Books, a division of Hearst Corp., 1990

[21] Doenitz, Karl, *Memoirs Ten Years and Ten Days*, Da Capo Press, Inc., 1997, pp. 440–441

[22] *Chronik Der Deutschen*, Chronik Verlag, Germany, 1983, p. 926

[23] None of these witnesses survived.

[24] *Chronik Der Deutschen*, Chronik Verlag, Germany, 1983, p. 926

[25] Some facsimiles are, especially lithographed, very deceptive. This is true of many Hitler's documents. From: K. W. Rendell, *History Comes To Life*, University of Oklahome Press, 1995.

[26] Contrary to the real Hitler who used to carry a caliber 6.35 mm pistol, the double had a larger 7. 65 mm pistol in a holster.

[27] The comments refer to the real Hitler before April 22, the date of his disappearance and for the double thereafter.

[28] He even wore darned socks.

THE "SUICIDE" ON APRIL 30

We now come to the more puzzling aspects of this apparent murder and suicide story.

Here we encounter many, totally different, versions of what was supposed to have happened on that day.

The official version widely accepted by the then Western powers and media outlets was based on a report, later published as a book by Trevor-Roper.[1] He was later appointed a Regis Professor of History at Oxford and during the war was working for British Intelligence (MI5). This man was chosen by the British Government to conduct an investigation in response to Russian statements that Hitler was alive in Spain, or even in Westphalia (a German State in the British occupied zone). Western newspapers, at that time, carried those stories widely.

This was bad public relations for the United States and Great Britain at the end of the terrible war and it had to be counteracted as soon as possible. Trevor-Roper, in his Royal Air Force Major's uniform, was given all access to prisoners of war and other potential witnesses (except to the most important ones, now in Russian prisons). He was only briefly allowed to see the bunker (which he found utterly neglected with two or three inches of water covering the floor, concluding: there had been no forensic investigation done by the Russians). He also had the full co-operation of the U.S. military.

In his press release on November 1, 1945 and after little more than four weeks of investigation, "this droll man, and a master of tart understatement", as O'Donnell called him, managed to convince the international press that Hitler was indeed dead despite the absence of any corpse.

Here is the gist of his report:[2]

> Available evidence sifted by British Intelligence and based largely on eyewitness accounts shows – as conclusive as possible without

bodies – that Hitler and Eva Braun died shortly after 2:30 A.M. (most say it was 3:30 P.M.) on April 30, 1945, in the bunker of the Reich Chancellery, their bodies being buried just outside the bunker.

On the evening of April 29 Hitler married Eva Braun, the ceremony being performed by an official from the Propaganda Ministry in a small conference room in the bunker....

At about 2:30 A.M. on April 30 Hitler said goodbye to about twenty people, about ten of them women, whom he summoned from the old bunker in the Old and New Reich Chancelleries. He shook hands with the women and spoke to most of them.

On the same day, at about 2:30 P.M., though the time is uncertain, orders were sent to the transport office requiring the immediate dispatch to the bunker of 200 liters of petrol. Between 160 and 180 liters of petrol were collected and deposited in the garden just outside the emergency exit of the bunker. At about the same time Hitler and Eva Braun made their last appearance alive. They went around the bunker and shook hands with their immediate entourage and retired to their own apartments, where they both committed suicide, Hitler shooting himself, apparently through the mouth, Eva Braun apparently, by taking poison, though she was supplied with a revolver.

After the suicide the bodies were taken into the garden, just outside the bunker, by Goebbels, Bormann, perhaps Colonel Stumpfegger, and one or two others. Hitler was wrapped in a blanket, presumably because he was bloody.

The bodies were placed side-by-side in the garden about three yards from the emergency exit of the bunker and drenched with petrol. Because of the shelling, the party withdrew under the shelter of the emergency exit and a petrol-soaked and lighted rag was thrown on the bodies, which at once caught fire. The burial party then stood at attention, gave the Hitler salute, and retired.

From there on, the evidence is more circumstantial. How often the bodies were re-soaked or how long they burned, is not known. One witness was informed that they burned until nothing was left; more probably they were charred until they were unrecognizable, and the bones broken up and probably buried.

The above evidence is not complete, but it is positive, circumstantial, consistent, and independent. There is no evidence whatever to support any theories that have been circulated and which presuppose that Hitler is still alive. All such stories which have been reported have been investigated and have been found

quite baseless; and some of them have been admitted by their
authors to have been pure fabrication.

All in all this was not a bad report aside from some minor errors
(such as that E. Braun had a pistol instead of a revolver, the
timing of the marriage was wrong, etc.), and considering that he
could not interview the key witnesses being either dead or
Russian prisoners.

However, what is remarkable about this story is the part of
Hitler and Eva Braun sitting in his studio and Hitler shooting
himself while Eva Braun apparently poisoned herself. Where did
this come from? The only witnesses Trevor-Roper had to
interview about what went on *inside* the Bunker where Erich
Kempka, Gerda Christian and Else Krueger (Bormann's
secretary). Kempka admitted sheepishly to O'Donnell in 1974
that he was not even among those present in the corridor at that
critical moment! Neither were Else Krueger and G. Christian. It
is more than likely that all three persons were at that time in their
living quarters, which were located on the upstairs bunker floor.
Therefore, they could only tell what they might have snapped up
from conversations with the valet, Linge, and others. For
example, G. Christian, one of Hitler's secretaries, stated that she
learned of the Hitler suicide from Linge and G. Junge, another
secretary, learned it from SS Major Guensche. Junge further
claimed that, according to Guensche, Hitler's ashes were collected
in a box that was given to Youth Leader, Arthur Axmann. This
was of course denied by Axmann.

However, it fits with the Nordon report stating that no bodies
of Hitler, or of Eva Braun, were ever found.[3] Please consider also
that the whole Trevor-Roper report stated quite clearly, at the
onset, that the investigation was made "without bodies", that is
without physical evidence whatsoever! It seems that Trevor-
Roper, at one time, was told by the Russians that they had found
two burned bodies (See page 77, "What the Russians discovered"),
except that this conflicts with one of Trevor-Roper's witnesses'
statement "that the bodies were burned till nothing was left".

Yet this story of the dual suicide in Hitler's anteroom
nevertheless became something of a historical fact thereafter.
Even the Russians later adopted this version, although with some

minor variation, in 1968, long after Stalin's death. Stalin, the Russian leader at that time, quite likely knew of Hitler's escape. However, Trevor-Roper, being annoyed that the Russians would not accept his own findings, later dismissed Stalin insistence that Hitler was alive, by stating: "Stalin had decided to suppress the documents and falsify the evidence in 1945." Maybe out of wounded pride he further argued: "Either (Stalin) was genuinely unconvinced by the Russian inquiry, or he deliberately falsified its results for political purposes." Trevor Roper, as did many others, also suspected the accuracy of the Russian autopsy report on the two "burned" corpses. Yet, despite his misgivings, he choose to include the Russian version of what happened to Hitler in his third book. A grateful Great Britain made Trevor-Roper a peer of the realm.

Trevor-Roper's account of the carrying of the bodies up to the garden, and the following cremation, was apparently based on the testimony of a Hermann Karnau, who was supposed to have been an SS member (he was a police detective).

There is another major flaw to the story. Kempka, Hitler's chief chauffeur, told O'Connell in 1973, that when he was requested to supply petrol on April 30, he said there was insufficient petrol left.[4] He only could obtain about 180 liters. Still, he had sufficient petrol the following day (May 1st) to burn the bodies of Joseph and Magda Goebbels to near cinders. However, if one considers that Mueller's plan was to bury the unburned body of Hitler's double, in order to be found by the Russian troops, then Kempka's statements that there was only sufficient gasoline to burn two bodies (that of Dr. and Mrs. Goebbels) make sense. However, all these stories have to be taken with a grain of salt. For example, Kempka explained apparent contradictions in his story by saying, "Back in forty-five to save my own skin, I told American and British interrogators just about anything or everything they wanted to hear."

Here is another version of the events, according to Toland:

> Guensche called Kempka that he needed two hundred liters of gasoline. Impossible, replied Kempka all the gas is buried in the Tiergarten and we cannot get it due to artillery fire. At 3:30 P.M., on April 30, 1945, Hitler picked up a Walther pistol. He was

alone in the anteroom[5] of his quarters with Eva Braun. She was already dead. She was on a couch slumped over the armrest, poisoned. A second Walther lay on the red carpet, unfired. Hitler sat at a table. He put the pistol barrel in his mouth and fired. In the conference room, Bormann, Guensche and Linge heard the shot. They hesitated momentarily, then broke into Hitler's anteroom.

This version was apparently based on Kempka's original testimony in 1945, later retracted by Kempka. This was also disputed by Arthur Axmann who stated: "I was standing right there, as close to the door as possible, but I certainly heard no shot." Kempka also told his first American and British interrogators in 1945 that, when looking into the suicide room, he saw both corpses and he said: "….it was clear to me that the Fuehrer and Miss Braun shot themselves," and later: "While carrying Miss Braun upstairs I saw blood trickling out of her breast." This apparent contradiction in Kempka's testimonies was also noticed by the *London Times,* stating that Kempka's statement: "…left the mystery of Hitler's end as undecided as before."

The alert reader will have by now discovered that each of the surviving key players in this drama, Linge, Kempka, Rattenhuber and Guensche told a different story about what was supposed to have happened regarding the suicide! The question is why? Perhaps it was too hectic, or there was insufficient time to rehearse their story sufficiently for a foolproof cover-up.

Another version, pieced together by O'Donnell and based primarily on information given by Prof. Haase to Dr. Schenck, went as follows:

Hitler sat down on the left-hand corner of the narrow sofa. Next he took out of his tunic pocket two poison vials. One he placed on the table between the pistol and a vase. The other he put into his mouth. His bride, Eva, was seated in the other corner of the blue and white sofa. Eva put a poison capsule in her mouth. She apparently then bit into the capsule. Hitler must have put the muzzle of his black Walter directly to his graying left temple, right angle at eyebrow level. He then squeezed the trigger and simultaneously bit into his capsule.

Heinz Linge was supposed to wait ten minutes before opening the door to Hitler's anteroom, instead he dashed up the stairs and was the first to enter.

Note, if anything Linge would have dashed *downstairs*, since Hitler's rooms were on the lower level of the two-story bunker.

There are several questionable aspects to this story. First, it is based on hearsay; Prof. Haase was not present when this apparent suicide happened. Secondly, it would have been nearly impossible to simultaneously bite the poison and to shoot oneself. Thirdly, if no shot was heard (see previous testimony and the fact that the room was closed by a heavy metal door) how did Linge know that the suicide did happen before he entered the room? Finally, here we have Hitler shot in the temple instead of through the mouth! Another absurdity about part of Linge's and Schenck's testimony was "that Hitler shot himself with his left hand in his left temple".

First of all, Hitler was right-handed and secondly, his left hand usually trembled.

What is the strangest part of all these stories is the fact that the above witnesses could describe exactly what was happening inside a room whose soundproof door was closed and where the only occupants were dead! We must therefore conclude that at least part of this testimony is pure speculation.

Rattenhuber later told his Russian captors that Hitler had been given the "coup de grace" with his own pistol, in case the poison did not work. This testimony conflicted with Linge and Guensche's testimony to the Russians and, as a result, he got the latter two into a lot of trouble. Yet Rattenhuber's story may have well applied to the killing of Hitler's double.

James O'Donnell admitted, that it is possible, that Linge and Guensche might have been lying.[6] That is of course a distinct possibility since according to the SS General Mueller, at least Linge was a co-conspirator in the apparent murder of Hitler's double as we will see later. As such he must have been very careful in all of his testimonies in order to avoid being prosecuted later in court for being an accessory to the murder of Hitler's double.

The East German writer, Rosanow, in 1956 published a book that apparently was heavily censored by Moscow, stating that:

"Hitler shot himself in the mouth, Eva Braun had taken poison."
He also showed the familiar photo of the (on May 2 discovered
and well preserved) corpse of the Hitler double, proclaiming it
was the real thing. This despite the official Russian version that
Hitler's body was partly burned!

In his book *Hitler, The Survival Myth*[7], Donald McHale gives
yet another version quoting Robert Waite an American historian:

> Linge had discovered Hitler in the death room with his hands
> folded carefully in his lap. The Fuehrer took poison and then
> someone, but not Linge or Guensche, as the Russians assumed,
> had shot him. He surmised that it was Eva Braun who shot
> Hitler. This happened at 3:30 on the afternoon of April 30.

Now we come to the latest version of what happened. This is the
result of Anton Joachimsthaler's well researched book[8] which was
published only in 1995.

> After evaluating all the testimony and photographs, and after
> considering all the known circumstances *in situ*, Hitler (on the
> right) and Eva Braun (on the left) were sitting on the narrow, 1.70
> m long sofa in Hitler's living room-cum-office before the suicide.
>
> Eva Braun-Hitler then bit down on the prussic acid ampoule
> and probably fell over on to Hitler sideways. Subsequently Hitler
> lifted his pistol to his temple and pulled the trigger. After the shot
> his body then remained seated between Eva Braun and the
> armrest of the sofa with the head canted slightly forward to the
> right. The dropping right arm let go of the pistol, which fell to
> the floor. With the shot into the temple, blood dripped onto the
> armrest of the sofa and then flowed in a copious amount on to
> the rug in front of the armrest of the sofa.
>
> There can be no reasonable doubt that the bodies were those
> of Adolf Hitler and Eva Braun, given the witness accounts which
> agree on all major points. The bodies were seen by Linge,
> Guensche and Axmann. Kemka, Hofbeck, Schneider and
> Mansfield credibly testified that the bodies carried out of the
> office into the garden for cremation were those of Adolf Hitler
> and Eva Braun. Unmistakable identification was possible because
> Hitler's head was partially uncovered and his lower limbs with
> the black trousers, black socks and shoes were visible. Eva Braun-
> Hitler's corpse was uncovered.

What shall we make of this? On the surface it all sounds very plausible. However, if we assume that the person that was shot in reality was Hitler's double then the testimony by Linge, Guensche and Axmann make sense (including the continuing contradictions in detail). Remember these witnesses were sworn (according to Russian accounts) to pretend that the real Hitler committed suicide. The additional fact is that the Russians found this double, who was most likely shot inside the bunker.

We don't know who the so-called Eva Braun was but remember there was another testimony that her body too was covered. As to the *credible* testimony of the other witnesses, that carried Hitler upstairs to the garden, we know that only Hitler's forehead and not his face, was uncovered. Consider also the great similarity in appearance between Hitler and his double. Finally we have the pants, the socks and the shoes. Any forensic expert will laugh at this kind of "positive identification". After all, the first thing you do with a double is, to dress him in a like manner! Another point to remember is if Linge or Guensche tells everybody in the bunker that Hitler just committed suicide then why should not the other personnel in the bunker believe this and automatically assume that the corpse being carried out was the real Hitler. While there were traces of blood on the armrest of the sofa in Hitler's living room, there was also a large single pool of blood (10 to 12 inches in diameter) on the red carpet at least two feet in front of the sofa. This could not possibly have come from the blood trickling down the armrest.

Here is my final argument: Even though the blood-stained carped from Hitler's office was burned in the garden on orders from Linge, there was (according to photographs) sufficient blood on the arm rest of the sofa to allow for a blood sample. The results of such tests were never published, if the Russians ever indeed took blood samples. The likely reason: The blood was not Adolf Hitler's type.[9]

The official history of the U.S. Army states flatly; "Hitler shot himself with a pistol."

So, if you are confused, you are not alone. We may certainly assume that there was one dead Fuehrer that afternoon. But it must have been Hitler's double. His was the *only* corpse found.

According to Mueller's testimony, the double was first drugged and then shot in the forehead using a Walther 7.65 mm pistol.

Was the story of the "official" suicide in Hitler's anteroom true? Most likely *not* since, as we can see, it was based primarily on later recanted testimony and on a lot of hearsay. It is also uncorroborated by any forensic evidence, and, most importantly the absence of any corpses!

In any case, it would have been difficult to conduct a thorough forensic investigation of the rooms in question since the Bunker was looted by Russian troops prior to the 2 P.M. arrival of the Russian intelligence troops. (It was reported that Russian women soldiers brandished black-laced bras of Eva Braun). Besides, there are reports that the Bunker hallways were partly flooded due to the shut-off of the utilities. Also, no bullets were ever found. We also have to consider the report from General Mohnke that he ordered Hitler's study to be burned by Captain Schwaegermann on May 1. The room was set aflame with gasoline, according to the testimony of the technician Hentschel who was the last to leave the bunker. Hentschel further stated the steel door to the study was red hot and the rubber seals melted. Such a hot fire could have destroyed at least some of the evidence, at least in the anteroom.

A recent article by Ada Petrova and Peter Watson in the *Washington Post*, which was supposed to be the "full story" about Hitler's death, with new "evidence from secret Russian Archives", does not shed any more light on the story, but lists a number of fake sightings of Hitler among other known details. They stated that Trevor-Roper gave a lot of credence in his story to the existence of the Hitler testament and his wedding contract with Eva Braun (this wedding ceremony did not have one single surviving witness!). The authors stated: "The fact of the marriage tends to confirm the psychological portrait Trevor-Roper was putting together". (If you don't have enough facts, you substitute psychology). While the authors of this article admit that "the Trevor-Roper account was necessarily incomplete and that there were many gaps to be filled in", they concluded that the book published by Trevor-Roper in 1947 "by rights ought to have solved the mystery and once and for all, to have killed speculation for ever. It was meticulously researched, well written and by and large convincing. *But among several points left unresolved, one all-important matter remaining a mystery.*" [Emphasis added by this author]. Unfortunately the authors did not elaborate which matter remained a mystery.[10]

Returning to the surviving witnesses, please remember, everybody was under tremendous pressure in those days. There was constant bombardment above, stifling air inside the bunker, everybody was very scared of the Russian troops and concerned for their personal future. It must have been terrifying. As a result, there was heavy alcohol consumption.

On top of this there later was the very tough interrogation by Russian and Western intelligence officers. Each one trying to make the witness confirm his own pet theories and political agendas. For example, Guensche underwent hundreds of hours of interrogation in a Moscow prison and he was promised instant release if he would change his story towards the ideas of their captors of what happened in Berlin. Finally, there was the fear of giving incriminating evidence and possible later prosecution of some for apparently murdering Hitler's double.

Notes

[1] Trevor-Roper, Hugh, *Last Days Of Hitler*, third edition, *The Times*, London, 1946

[2] O'Donnell, James, *The Bunker*, DA CAPO Press, 1978

[3] Brown, Anthony Cave, *The Last Hero*, *Wild Bill Donovan*, Vintage Books, a division of Random House, 1984

[4] Toland, John, *The Last Days*, A Bantam Book/ Randam House, Inc., 1967

[5] According to other witnesses, the suicide happened in Hitler's study or living room, next to his anteroom.

[6] O'Donnell, James, *The Bunker*, DA CAPO Press, 1978

[7] McKale, Donald M., *Hitler The Survival Myth*, Cooper Square Press, 1981

[8] Joachimsthaler, Anton, *The Last Days Of Hitler*, Cassell & Co., London, 1995

[9] Irving, David, *Hitler's War*, Avon Books, a division of Hearst Corp., 1990

[10] Petrova, Ada and Watson, Peter, "The Death Of Hitler", *Washington Post*, 7–6–03

HITLER'S LAST WILL

Part of the clue to the puzzle of Hitler's escape may be found in his political testament.

Here is the official story: Hitler during the night of April 28, and following the midnight wedding reception supposedly wrote both a political and one personal testament, typed in three sets, which he then supposedly signed at 4 o'clock on the morning of April 29, 1945.

As we mentioned previously, Traudel Junge one of the two remaining Hitler secretaries, stated that the will was dictated to her from notes. She then wrote what was dictated down in shorthand and later transcribed the text onto her typewriter. This was the first time this ever happened to her, since the real Hitler always dictated to her while she was typing and without using any notes! All this, the notes and the style of dictation again points to the presence of a double. According to Ian Kershaw while the will was being dictated, "Goebbels- who together with Bormann, kept bringing Fraulein Junge the names of further ministers for typing in the list (the proposed members of a new government requested in the testament)."[1] The real Hitler would never have tolerated such interference!

The personal part of the will was rather short and discussed his marriage and his appreciation that "this girl entered this city, already besieged, of her own free will, in order to share my fate with me. At her request she is joining me in death as my wife..." Hitler also appointed Martin Bormann as executor of his will and requested him to assist his relatives and former secretaries.[2]

It ended by saying: "I myself, and my wife, chose death to escape the disgrace of removal or surrender. It is our desire to be burned at once at the place in which I have performed the greater part of my daily work in the course of twelve years of service to my people."

While this personal testament suggests suicide, it actually does not say so. It also pre-supposes that Martin Bormann would have been able to escape from Berlin, how else could he perform his function as executor of his will? Finally, Hitler ordered that his body should be burned. This last statement fitted well with the agreed upon cover story of the dual suicide with subsequent cremation.

The political will is much longer and a part of the text accused the international Jewry and its accomplices for ruining the cities and monuments. It begins:

> It is not true that I or anybody in Germany wanted war back in 1939. It was desired and provoked solely by those international politicians who either come from Jewish stock or are agents of Jewish interests. After all my offers of disarmament, posterity simply cannot pin any blame for this war on me…
>
> After a struggle of six long years, which in spite of many setbacks will one day be recorded in our history books as the most glorious and valiant manifestation of the nation's will to live.
>
> I cannot abandon this city, which is the German capital. Since we no longer have sufficient military forces to withstand enemy attacks on this city, and since our own resistance will be gradually exhausted fighting an army of blind automata, it is my desire to share the same fate that millions of other Germans have accepted and to remain here in this city…

He also reviewed the last twenty-five years of his struggle and its justification. However what is more important to our story is, that he named Admiral Doenitz as his successor in the post of President and he appointed Joseph Goebbels as Chancellor of the Reich. This portion of the will was faithfully passed on to Doenitz during the following day by Bormann in the form of a telegram. Otherwise this testament expressed exactly Hitler's political outlook that he had years ago. Assuming that he was the author of this document, then he simply tried to justify himself and his actions to the coming generations of Germans.

In this testament he hinted at dying in Berlin but left the method of his death open. However, reading this portion of his will and finding the dead double could then convince the Russians that he, Hitler, was indeed dead.

The trouble with this will is, that it could not have been written by Hitler's double, the only "Fuehrer" left in the bunker. First of all, there was little more than three hours time between the break-up of the fake wedding party (the ceremony itself did not end till after midnight, followed by a reception). It would certainly take more than an hour just to dictate this lengthy, and for Hitler quite important document, then it had to be transcribed from shorthand and finally edited, then, two additional copies had to be typed too. Finally, the double was not that familiar with the phraseology and the historical references made in the political will.

It was suggested by O'Donnell that the document was dictated using notes written by Dr. Joseph Goebbels.[3] That could not be true either, why would Goebbels write a will to appoint himself Chancellor and then two days later commit suicide?

This brings us to the most puzzling part and that is: Why did the real Hitler write a will on April 29, to name Goebbels Chancellor knowing quite well (as everybody else in the bunker knew) that Goebbels planned to commit suicide together with his family, which he did two days later?

According to O'Donnell, Goebbels confided in mid February 1945, to his aid, Lieutenant von Oven: "Neither my wife nor a single one of my offspring will be among the survivors of the coming debacle."[4] This clearly shows that the suicide was planned for a long time.

There are only two logical explanations for these contradictions. First, that the will was a forgery. One fact that speaks for this assumption is that the last two paragraphs alone, containing sixty words, include no less than six major spelling and grammatical errors. These errors are of such a nature, that it appears a person made them who was not too familiar with the German language, a foreigner for example. It is hard to believe that Hitler's remaining secretaries, who both had many years of experience, could have made such blatant spelling errors. It is even more unlikely that Hitler would have signed such a document that had all the hallmarks of being typed by an eighth grader. As any dealer in autographs knows there are many forgeries on the market that are so well done that they easily pass as originals. Furthermore, intelligence services have plenty of specialists and equipment to produce false passports and other documents. Of interest, in this connection, is the report that Ernst Kaltenbrunner the former head of the German State Security Service was captured in May 1945, in the Austrian Alps. He was supposed to have with him one copy each of Hitler's personal and his political testament. The strange part about this story is, that Colonel von Buelow, who gave it to Field Marshall Keitel, then in northern Germany, carried the only set that got out of Berlin! General Mohnke taped it to his body then carried the other two sets of testaments out of the bunker. The Russians later captured Mohnke, on May 2, 1945. It therefore can be assumed that these only remaining two copies of the wills are kept in Russia. However, the assumption of a forgery was supposedly disproved by a U.S. forensic investigation of at least one of the original three copies that were sent out of the bunker.

The second explanation is that the two documents were actually written some time *before* April 22, 1945, when Hitler thought he still could talk Goebbels out of the suicide idea (he certainly tried). Note also that Dr. Goebbels and his wife where listed as passengers on the original flight manifest to Barcelona, Spain. This of course also agrees with the story SS General Mueller told his interrogators namely that the will and the marriage contract were written and signed prior to Hitler's departure.[5] All Goebbels or Bormann had to do was to have the witnesses sign the documents. The double probably was not even asked to attend the 4 A.M. signing, at that time he was probably sound asleep.

This, of course still does not explain the many typing errors in the document.

Other than a forgery they can only be explained by the extreme pressure, the late hours, perhaps alcohol consumption and the knowledge that the real Hitler was no longer there. Then again, the published document, showing these errors, might have been a poor copy of the original wills. Yet we may conclude, that these puzzling aspects of the wills could be considered as additional evidence that the real Hitler was not in the bunker at the given time frame.

Here I would like to add some comments on, what I believe was a "fake" wedding. The ceremony was to have started shortly before midnight on April 28 and was said to have taken place in the small anteroom next to Hitler's living quarters. This anteroom was sealed by a soundproof steel door. A local notary, Walter Wagner, officiated. Besides him there were only Hitler, Eva Braun, Goebbels and Bormann in the room. All of these were dead within the next three days; the poor Wagner was shot dead within the next hour!

Yet, we are reading in Fest's book,[6] description of the whole speech by Wagner, and I quote:

> I come here to the solemn act of matrimony. In the presence of the above-mentioned witnesses... I ask you, My Leader, Adolf Hitler, whether you are willing to enter into matrimony with Miss Eva Braun. If such is the case, I ask you to reply "yes".
>
> Herewith I ask you, Eva Braun, whether you are willing to enter into matrimony with my Leader, Adolf Hitler. If such is the case, I ask you to reply "yes".
>
> Now, since both these engaged persons have stated their willingness to enter into matrimony, I hereby declare the marriage valid before the law.

This verbatim quote of what went on behind closed doors is, to put it mildly, very puzzling, since there is not a single surviving witness to this scene. Again, we can assume that the only thing that went on here was that the wedding document, pre-signed by Hitler and Braun was then countersigned and dated by Wagner and the witnesses.

This of course explains why SS General Mueller had Wagner liquidated within half an hour after the "ceremony".[7] If the Russians would have caught him he might have talked and exposed the whole "double spiel" by admitting that there was no "real" Hitler and for that matter no "real" Eva Braun in the bunker. Otherwise the wedding nicely reinforced the desired impression that Hitler and Braun were still in the bunker.

Following the "wedding" there was a reception in a small conference room. According to Fest, it was attended by the "newly weds", the secretaries, the adjutants and Miss Manzialy,

Hitler's dietary cook for many years. This small party, attended only by low level bunker insiders, had drinks and reminisced. It would have been interesting to know whether the double drank champagne? If he did, then his behavior would have been quite a departure from the habit of the teetotaler Hitler. Also strange is why were Bormann and Goebbels absent on such a special occasion? We will never know. However, one explanation could be that it was beyond their dignity to socially associate themselves which such a small fish as "Hitler" the double. Besides, Goebbels was busy that night, writing, dictating and adding his personal testament to that of Hitler's.

As to Miss. Manzialy, Hitler's cook, she also disappeared shortly thereafter. Rumor has it that she poisoned herself on May 2, 1945.[8]

---

[1] Kershaw, Ian, *Hitler*, W. W. Norton & Co., 2000

[2] Fest, Joachim C., *Hitler*, A Harvest Book. Harcourt, Inc., 1973

[3] O'Donnell, James, *The Bunker*, DA CAPO Press, 1978

[4] O'Donnell, James, *The Bunker*, DA CAPO Press, 1978

[5] Douglas, Gregory, *Gestapo Chief, The 1948 Interrogation of Heinrich Mueller*, James Bender Publishing, 1995

[6] Fest, Joachim C., *Hitler*, A Harvest Book. Harcourt, Inc., 1973

[7] Douglas, Gregory, *Gestapo Chief, The 1948 Interrogation of Heinrich Mueller*, James Bender Publishing, 1995

[8] Overy, Richard, *Interrogations, The Nazis In Allied Hands 1945*, Penguin Putnam, Inc., 2001

A VIKING FUNERAL

If, for the sake of discussion, one would assume that the participants in the conspiracy surrounding Hitler's escape wanted to cover his tracks, then one way this could be accomplished is by pretending that Hitler and Eva Braun committed suicide, assuming the ruse with the double did not work. Well and good, but how about the bodies? Since both disappeared, there could have been no corpses. Here is a way out of this dilemma: you pretend that the corpses were burned beyond recognition, or, disintegrated into ashes.

While we have no proof that this is what really happened (the conspirators would not admit to it in public) the fact that the Russians could not find corpses of either the real Hitler nor Eva Braun gives this explanation more credence.

Anton Joachimsthaler quoted in his book the text of a report, supposedly issued in May 1946, by a high level Russian Task Force formed to investigate Hitler's fate, saying,

> Not a trace was found of the bodies of Hitler and Eva Braun. Nor was there any trace of the petrol-drenched grave in which the bodies of Hitler and his companion were allegedly cremated according to a statement by some of the witnesses. Some witnesses have now confessed to swearing an oath to Hitler that if they were captured, they would claim to have seen Hitler's and Eva Braun's bodies being burned on a pyre in the bunker garden. All the witnesses have now admitted to the investigating committee that they did not see a pyre, nor Hitler's or Eva Braun's bodies.
>
> It has been determined that Hitler attempted to cover his tracks with the help of false witnesses.
>
> There is irrefutable evidence that a small aeroplane took off from the Tiergarten in the direction of Hamburg. It is known that there were three men and one woman on board. It is also determined that a large submarine left Hamburg harbor before British forces arrived. On board were mysterious people, including one woman.[1]

This makes a lot of sense when looking at the overall evidence. For example, the first part agrees in all essential points with the Nordon report shown elsewhere.[2] Joachimsthaler[3] agrees that the Russians never found a body of Hitler, burned or otherwise, yet he dismisses all Russian statements by Stalin and other Russian officers as being untrue.[4] Why not at least consider that they could be true?

The question is why should the Russians lie? What motives would they have? Why not allow that the only three surviving main witnesses to the Viking funeral namely Guensche, Linge, and Kempka, all being fanatical Nazis, were lying? To me this seems much more logical.

Concerning the last paragraph of the above report, the Russian generals are surely guessing about the means of Hitler's escape, having no hard evidence to rely upon, in contrast to the physical evidence in the bunker garden. However, it is true that Hanna Reitsch and General Ritter von Greim, SS General Mueller, and finally Hitler's adjutant, von Below, all flew out of Berlin in small planes during the last days of April 1945. This certainly accounts for the last part of the Russian report.

Now let's start to analyze the well-reported and most widely believed story of the funeral as told by Trevor-Roper and others:

> After the suicide the bodies were taken into the garden, just outside the bunker, by Goebbels, Bormann, perhaps Colonel Stumpfegger, and one or two others. Hitler was wrapped in a blanket, presumably because he was bloody. According to Trevor-Roper,
>
>> the bodies were placed side-by-side in the garden about three yards from the emergency exit of the bunker and drenched with petrol. Because of the shelling, the party withdrew under the shelter of the emergency exit and a petrol-soaked and lighted rag was thrown on the bodies, which at once caught fire. The burial party then stood at attention, gave the Hitler salute, and retired.
>
> From there on the evidence is more circumstantial. How often the bodies were re-soaked or how long they burned, is not known. One witness was informed that they burned until nothing was left; more probably they were charred until they were unrecognizable, and the bones broken up and probably buried.

Hitler and Eva Brain with "Blondie" and her dog at Berchtesgaden in June of 1942.

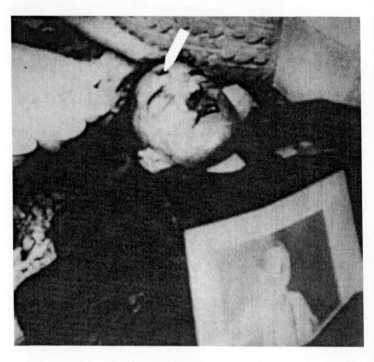

Official Soviet photograph of the dead double of Hitler, taken in the Court of Honor of the Reich Chancellery, after exhumation on May 2nd 1945. Note bullet hole in forehead (see arrow).

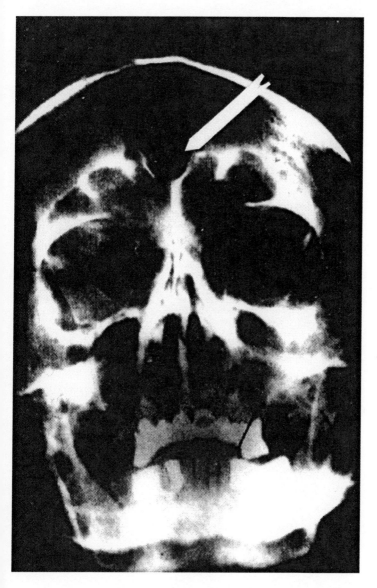

X-ray supposed to be taken of Hitler's head in September 1944. Note circular hole (see arrow) between the two upper sinuses matching the bullet hole in the forehead of Hitler's double.

Photo taken of the bunker exit to the garden. Note the remaining wood scaffolding and the loose wood planks on the sandy garden showing no signs of a raging fire. The circle marks the spot where Hitler and Eva Braun were supposedly cremated. The observation tower is shown on the right.

As to the funeral party which carried the bodies into the garden and then burned the bodies of the alleged Adolf Hitler and Eva Braun, there were according to John Toland[5] and Trevor-Roper:[6] Linge, Goebbels, Bormann, Dr. Stumpfegger, Guensche and Kempka.

From the outset let's remember there were only *three* surviving witnesses to this alleged happening: Linge, Kempka, and Guensche. All others did not survive the end in the bunker! So this makes for a very limited verification, especially if each of those witnesses were sworn to plant this particular story.[7]

The other trouble with this version is the fact that it relies also on hearsay and limited observations from SS guards who supposedly witnessed the scene from afar but some of whom could not be located later on.[8] I will analyze their testimony later.

Out of the six present at the funeral, only three survived the end of the war and as we know, both Linge and Guensche stated later (after their return from Russia) that they did not see the faces of the corpses. As to Kempka's testimony, he said in 1945 that he was unable to identify the body of Hitler when he helped carry the corpse of a male out of the bunker.[9] "The body was covered by a blanket and I could only see black trousers and black shoes similar to the one's worn by Hitler." This again contradicts his earlier story that he, Kempka, carried *Eva Braun* upstairs. He contradicted himself again by saying first Eva Braun was wearing a black dress, then later on, that she was wearing a blue dress with white trimmings. Similar contradictions appear in the testimony of other witnesses. For example, Mohnke stated that Eva Braun was covered in a blanket and was wearing no shoes, while Kemka stated there was no blanket, and SS Guard Karnau saw Eva Braun's familiar black shoes protruding from under the blanket *after the fire had been burning for ten minutes!*

Anton Joachimsthaler gives the most detailed, recent (1995), and quite well researched account of the funeral in his book, *The Last Days Of Hitler.*[10] Quoting Linge as saying:

> We placed Hitler's body a short distance from the garden exit of the bunker. Immediately after this Guensche appeared with Eva Braun's body, which we placed next to that of Adolf Hitler. The bodies lay directly next to each other with the *feet pointing towards the bunker exit…*

Now contrast that with the testimony of Guensche who stated:

> I want to make it clear that the *heads were pointing in the direction of*
> *the bunker exit*, and that – seen from the bunker exit – Eva Braun
> was lying to Hitler's right.

Of further interest are the statements by Linge, Kempka and
Guensche that they did not complete the cremation but left this
task to some unnamed underlings. In other words, even though
they were ordered to make sure no traces remained of Hitler's
corpse, they did not bother to check if this was actually done. Very
curious indeed! Even Trevor-Roper remarked: "From this moment
on no one seems to have given a damn about the past or the two
bodies that were still sizzling in the garden."[11] This behavior makes
sense only if there were no burned bodies to start with.

As I mentioned before, we have testimony by the German
guards Karnau,[12] Mansfield, and Hofbeck on whose testimony
Trevor-Roper heavily relied.

Mansfield said that he observed the burning of two corpses
from the observation tower which was about 45 feet from the
cremation site. He further testified that heavy smoke obscured his
vision, since a wind was blowing in his direction. His vision was
further restricted by a slotted steel window. He stated that cans of
gasoline were thrown from the bunker exit and towards the
cremation site. From his position and under the given
circumstances, we can safely assume that he would not have been
able to identify anybody on the ground.

Let's see now what happened with the remains of the
supposedly burned corpses.

Here we have Kemka's statement that he buried the remains
in the garden and outside the wall of his apartment. Yet this
statement was obviously false since nothing there was found by
the Russians. Rattenhuber said that he gave no orders to bury the
remains. This leaves only Guensche who stated that he ordered
SS Officer Lindhoff to bury the remains in the garden.
Unfortunately Lindhoff was killed trying to escape from Berlin.
We should add to this that one of the SS guards, named Karnau,
stated on November 13, 1953 that he saw skeletons of both bodies
at around 5 P.M. on April 30, 1945. Yet he corrected his statement

on June 30, 1954 by saying: "I did *not* see any bones. What I found was a pile of ashes, which disintegrated when touched by my foot."[13] This makes more sense and he could have been seeing the remains of the red carpet that was burned and of the destroyed Hitler papers and belongings that were cleaned out of Hitler's rooms and which were burned that afternoon. As we all know, remains of burned paper disintegrates when touched.

A similar statement was made by the guard, Hofbeck, who testified that around 8 P.M. there was nothing left to see but flakes of ashes. These statements also agree with Hentschel's testimony that on May 2, he saw no remains in the garden other than the charred bodies of the Goebbels.

We may now look at the cremation itself. Gasoline in liquid form does not burn. Only gasoline vapor does, which requires heat. Since the soil was sandy, we can assume that a major portion of the gasoline disappeared into the ground and thus did not burn at all. Finally, we had testimony that gasoline cans were thrown from the bunker exit towards the corpses (a distance of about 3 meters or 9 feet). We can therefore again assume that a good deal of this liquid got wasted. Since the ground was fairly even and since a wind was blowing, it can further be assumed that a good deal of the flames (creating the heat) were blown sideways and away from the alleged corpses. Gasoline fires can reach temperatures between 1300 and 1500 degree Fahrenheit, not enough to melt gold[14] for example. Modern gas fired crematoria are much more efficient and can get as hot as 1800 degrees F, partly due to the surrounding walls providing extra heat from radiation, in contrast to an open fire. Yet even here, the bones of a person do not disintegrate into ashes, as Joachimsthaler[15] will have us believe, although extreme heat makes bones brittle.[16] Modern crematoria solve this problem with the remaining bones by grinding them up and by putting the resultant powder in an urn, which is then given to the bereaved. Aside from the photos of the only partially burned corpses of Joseph and Magda Goebbels, there are quite a few photos published of crematoria in German concentration camps. In almost all of these pictures we see skeletons either still in the oven, or in front of it, despite the fact that these were modern, gas fired devices.

While Joachimsthaler conceded that the Russians did not find the corpse of Hitler, he still insists that Hitler committed suicide. His theory was that there was in excess of 200 liters of gasoline burning for over two hours. This is hard to believe reading all the testimony about the extreme difficulty of obtaining gasoline. Then he further theorizes that the shrapnel from the Russian shelling and the use of Napalm made the bones disintegrate into ashes.[17]

The problem with this theory is that Napalm[18] was not used by the Russians in 1945. Furthermore there could not have been much shelling, since the bodies of the Goebbels couple, aside from the burning, showed no sign of damage. Furthermore, there was plenty of wood scaffolding next to the emergency exit and at the observation tower, due to unfinished construction. As photos taken shortly after the war show, there was hardly any damage at all to any of these, even of those planks on the ground (see attached photo).

Of course, if the Russians really found two burned corpses somewhere in a shallow grave, then we must accept the story of the funeral as being true. The question only is: since it could not been Hitler and his wife, who then were those dead persons?

We know for sure that at least two additional persons died from gun shot wounds inside the bunker: General Krebs, and Franz Schaedle.

Incidentally, according to Donald McKale, Johannes Hentschel the chief electrician for the Chancellery and the last person to leave the bunker testified that he saw only the charred bodies of Dr. and Mrs. Goebbels. Of Hitler, Eva Braun, and others from the last days he saw nothing.[19] This agrees with General Mueller's story that only Hitler's double, was buried in a shallow grave. This body was truly discovered on May 2, 1945 by Colonel Klimenko as we see later.[20] However, someone in the bunker decided to bury another corpse made out to look like Martin Bormann. Mueller apparently did not know this, since he left already on April 29. Klimenko too discovered this body, but this did not make the history books.

This brings up the interesting question: What happened to Hitler's double after he was shot (presumably inside the bunker) and who buried him? The fact that we have no testimony concerning this matter could only be explained by a) No interrogator ever asked that question of the surviving witnesses; or, b) The corpse that bystanders, such as Arthur Axmann, saw being carried upstairs, wrapped in a blanket with his face partly covered, was that of the double, instead of Hitler.

Well, my readers may rightly ask, if the corpse being carried upstairs and later buried in a shallow grave (instead of being burned) was that of Hitler's double, then what happened to Eva Braun? Unfortunately, we have no evidence of what really happened to her.

However, here is a plausible scenario. Remember, the young woman, the strawberry blonde (as Dr. Schenck described her) certainly was a substitute for the real Eva, provided by Mueller's security service. Remember, witnesses described that she arrived, together *with* the Goebbels family, at around 6 P.M. on April 22, 1945.[21] From thereon she played her part well, staying mostly in the background and seemingly avoiding all social contact with Hitler's double.

In the afternoon of the alleged suicide she probably played dead, for the benefit of all onlookers, and had her limp body carried upstairs by fellow conspirators Bormann, Guensche and Kempka. Once in the garden she most likely ran away to her own private residence. Some bystanders entering the living room afterwards sensed a strong smell of almonds (prussic acid) and therefore thought Eva Braun had poisoned herself. However this smell could have come from the mouth of Hitler's double, who, according to SS General Mueller was first poisoned and then shot.[22]

Notes

[1] Joachimsthaler, Anton, *The Last Days Of Hitler*, Cassell & Co., London, 1995, p. 24

[2] Brown, Anthony Cave, *The Last Hero*, *Wild Bill Donovan*, Vintage Books, a division of Random House, 1984

[3] Joachimsthaler, Anton, *The Last Days Of Hitler*, Cassell & Co., London, 1995, p. 252

[4] Joachimsthaler, Anton, *The Last Days Of Hitler*, Cassell & Co., London, 1995, p. 252

[5] Toland , John, *The Last Days*, A Bantam Book/ Randam House, Inc., 1967

[6] Trevor-Roper, Hugh, *Last Days Of Hitler*, third edition, *The Times*, London, 1946

[7] Joachimsthaler, Anton, *The Last Days Of Hitler*, Cassell & Co., London, 1995

[8] Douglas, Gregory, *Gestapo Chief, The 1948 Interrogation of Heinrich Mueller*, James Bender Publishing, 1995

[9] Kempka, Erich, *"Erklaerungen von Herrn Erich Kempka vom 20–6–45 und Ergaenzende Erklaerungen des Herrn Erich Kempka"* vom 4–7–45, given in German to the U.S. investigating officer Harry Palmer.

[10] Joachimsthaler, Anton, *The Last Days Of Hitler*, Cassell & Co., London, 1995, p. 192

[11] Trevor-Roper, Hugh, *Final Entries 1945 The Diaries Of Joseph Goebbels*, G. P. Putnam's Sons, 1978

[12] Even Ian Kershaw in his book *Hitler* commented on Hermann Karnau's testimony: "...like a number of witnesses in the bunker, he gave contradictory versions at different times."

[13] Joachimsthaler, Anton, *The Last Days Of Hitler*, Cassell & Co., London, 1995

[14] Gold has a melting point of 1945° F.

[15] Joachimsthaler, Anton, *The Last Days Of Hitler*, Cassell & Co., London, 1995

[16] Dix, J. and Calaluce, R., *Forensic Pathology*, CRC Press, LLC, 1998, p. 83

[17] Joachimsthaler, Anton, *The Last Days Of Hitler*, Cassell & Co., London, 1995

[18] Napalm, made by Dow Chemical Corp, is made of a mixture of aluminum salt and petroleum jelly. It was widely used in Vietnam.

[19] McKale, Donald M., *Hitler The Survival Myth*, Cooper Square Press, 1981

[20] Bezymenski, Lev, *The Death Of Adolf Hitler*, Michael Joseph, London, 1968

[21] The real Eva Braun moved into the Bunker on 4–15–1945 and left, apparently with Hitler on the 22 of April. She was last seen in the bunker around 5 P.M. on that day.

[22] Douglas, Gregory, *Gestapo Chief, The 1948 Interrogation of Heinrich Mueller*, James Bender Publishing, 1995

WHAT THE RUSSIANS DISCOVERED

The Russian generals in Berlin learned of the alleged suicide of the Fuehrer from General Krebs on May 1, 1945, while he was trying to negotiate the surrender of Berlin. When the negotiations failed, Krebs went back to the bunker and killed himself.

Russian female soldiers first entered Hitler's bunker at mid-morning on May 2, 1945, the day Berlin fell, and carried out some looting. In mid-afternoon the first five search teams (reporting to the NKVD, the Russian secret police) arrived led by Lieutenant Colonel Ivan Klimenko.[1]

He soon discovered the charred bodies of the Goebbels', which he rushed back to his headquarters. Later on a search team located one body, among others, in an old oak water tank (According to General Mueller, an empty, decorative stone pond in front of the new Chancellery. It was used as a temporary morgue for the field hospital located under the Chancellery. The corpse looking like Hitler was actually in a shallow grave next to the pond.)

The Russians then used the imprisoned German, Admiral Voss to identify this body. He declared that it was that of Adolf Hitler, even though the corpse wore darned socks. When Klimenko returned the next morning, he found the corpse prominently displayed in the main hall of the Reich Chancellery.

There is a widely published photograph (see attached photo) of this body showing a hatless person, looking very much like Hitler with a bullet hole in his forehead. This body was not burned at all. If it was the real Hitler, then the photo must have been taken inside the bunker, before he was carried out and burned (as the Russians and Trevor-Roper later claimed). To add to the confusion, some photos were published in reverse, possibly to misrepresent details. Some pictures show a blurred portrait of Hitler on top of the double's chest. No other photos were ever released by the Russians despite the extreme importance of this forensic investigation.

It is certain that this was the photo of the first Russian discovery and we can safely conclude that this corpse was that of Hitler's double. According to the "Nordon report" the Russians, too, later identified this corpse as being that of Hitler's double.

It was then ordered that the corpse of the double be cremated, but this process was interrupted on orders from Moscow. The partly cremated corpse was then flown to Russia.

Note: nothing in this official version of events indicated that a second double, that of Martin Bormann, was also found, as indicated in the Nordon[2] report.

Now comes a widely reported story that made it into the history books, but most likely was a hoax planted by the Russian Secret Service (NKVD) to cover up the fact that Hitler disappeared. It is contradicted by the Nordon[3] report, the statement by SS General Mueller[4] and later too contradicted by official Russian publications, printed after 1960.

Here goes the story:

A day went by (it is now May 3, 1945), but by then the Russian team had already decided that they had not found the real Hitler. The Moscow paper *Pravda* reported on May 3, "Hitler is not in Berlin".

This supposedly caused more frantic searches in the Garden with the result that one Private Churakov climbed into a nearby crater and discovered some legs. After more digging, they unearthed the bodies of a man, a woman and two dogs. Since this seemed not to be what the team was looking for, Klimenko ordered the corpses to be re-buried.

Finally on May 5th (probably after more frantic phone calls from headquarters) it clicked: "A man, a woman and dogs", that could be Hitler and Eva Braun!

Klimenko went back to the garden and had the bodies dug out again. He then had the badly burned corpses sent to the Field Hospital in Berlin-Buch. A special team, headed by Dr. Faust Sherovsky, was flown in from Moscow, the bodies were dissected and the first forensic autopsy was performed.

Looking at the partly burned body of the man, they found that the cranium was missing, the man had only one testicle and one seminal cord (Hitler had two). There were no fingerprints since the skin was burned and they discovered glass splinters from a thin-walled ampoule in his mouth. From this they determined "instant death by cyanide poisoning".

This was not much to go on. Also all the testimony of witnesses, at that time, pointed to death by shooting! Was this a false trail? The Russian Marshall Sokolovsky was very eager to prove that Hitler was dead, since it was quite embarrassing for him should the Fuehrer have escaped from under his nose. So he spared no effort in the investigation.

Fortunately, the corpse had some teeth and bridgework left. So they removed the lower jaw and the bridge and started to hunt for Hitler's dentist and his dental records.

This was difficult, since the German SS General Mueller had all dental records together with Dr. H. Blaschke, Hitler's dentist, flown out of Berlin.[5]

Incidentally, these records were always kept at the Chancellery and not at Dr. Blaschke's office. Mueller's agents then destroyed the dental records as a precaution.[6] Dr. Blaschke himself was subsequently captured and held by the American Army but the Russians for some unknown reason refused to request his appearance in Berlin. Blaschke was subsequently asked by his U.S. CIC interrogators, to reconstruct Hitler's dental work, but thought he could not do it without his files, which, of course were never found. However Blaschke did give a description of Hitler's teeth[7] and I quote:

Edge to edge bite, 6 upper, 10 lowers only remaining natural teeth. Upper right: 1 Richtman crown, 3, 4, full gold crown, 2&5 dummies. Left: 1 three quarter gold crown, 2 Richtman crown, 3 full gold crown, 4 dummy. Single fixed gold bridge over all uppers. Lower right: 1, 2, 4 normal, 3. 5 full gold crown with lingual bar between. Left: 1 normal, 2 porcelain filling and a pical abscess, 3, 5, 8 full gold crowns, 4, 6, 7 dummies.

General Donovan sent this information by telegram on July 28, 1945 to the Russian NKVD General Fitin.[8] However, for some reason, it seems that the Russians never used this vital information. However, this is understandable, if we consider that the Russians simply did not have the real Hitler corpse.

Luckily, the Russians then discovered two dental technicians, Fritz Echtmann and Kaethe Heusemann. The Russians also claimed to have found dental records and crown work of Hitler in Blaschke's office (which can not be true, according to Mueller, (see above).

In any case, they asked the technicians to sketch the dental structure "from memory". It should here be noted that apparently, K. Heusemann only helped to treat Hitler once in November of 1944. She later was sent for eleven years to Siberia. This apparently was done to prevent her from publicly contradicting her alleged identification.

So much for the official Russian version of the "two burned corpses".

By May 15, 1945 the Russian military concluded that this burned corpse truly was that of Hitler. However, Stalin having been told already about two Hitlers (the unburned "double" and now the new, "burned" corpse) was still not convinced and told his generals they could be mistaken. As a matter of fact, Stalin, on May 26, 1945 told the U.S. Envoy Harry Hopkins in Moscow: "In my opinion Hitler is not dead but is hiding somewhere".[9]

As a result, the Russian Marshall Zhukov held a press conference on June 9, 1945 in which he said: "We could not identify the body of Hitler. I can say nothing definite about his fate. He could have flown away from Berlin at the very last moment." To that the Russian General Bezarin added: "He has disappeared somewhere in Europe, perhaps in Spain with Franco. He had the possibility of taking off and getting away." A few days later Zhukov told General Eisenhower privately, there was "no solid evidence of Hitler's death."

Thus the whole affair began to resolve around the Russian's military desire for "closure" and Marshall Stalin's justifiable suspicion that he had been hoodwinked.

According to Gregory Douglas,[10] on May 13, 1945, *Pravda*, the Soviet Newspaper, said, "Moscow has directed the senior officers of the Red Army in Berlin to discuss nothing about the situation in the Fuehrer bunker."

On July 17, 1945, at the Potsdam Conference, the Russian leader, Josef Stalin, told U.S. President Harry Truman and U.S. Secretary of State, Byrnes, over lunch that he thought that Hitler was still alive "in Spain or Argentina". A few days later, he repeated his story to Churchill and to Ernest Bevin, the new Foreign Secretary for Britain.[11] It is quite possible that Stalin got information about Hitler's escape directly from his "mole" inside

Hitler's headquarter. This mole transmitted, starting in 1942, all vital German military orders directly to Moscow, typically within twenty-four hours after Hitler issued them. In his book, Louis Kilzer makes a strong case that Martin Bormann was this mole.[12]

Even more revealing is a report written by a Russian high-ranking military commission, originally chaired by Marshall Zhukov (and later by secret service chief Beria) written around May 1946 and supposedly given to the U.S. Envoy Harry Hopkins. It reportedly stated:[13]

> Not a trace was found of the bodies of Hitler and Eva Braun. Nor was there any trace of the petrol-drenched grave in which the bodies of Hitler and his companion were allegedly cremated according to statements by some of the witnesses. Some witnesses have now confessed to swearing an oath to Hitler that, if they were captured, they would claim to have seen Hitler's and Eva Braun's bodies being burned on a pyre in the Chancellery garden.
>
> All these witnesses have now admitted to the investigating committee that they did not see a pyre, nor Hitler's or Eva Braun's bodies.
>
> It has been determined that Hitler attempted to cover his tracks with the help of witnesses.
>
> There is irrefutable evidence that a small aeroplane took of from the Tiergarten in the direction of Hamburg. It is known that there were three men and one woman aboard. It was also determined that a large submarine left Hamburg harbour before British troops arrived. On board were mysterious people, including a women.

While the first part of this statement, being based on physical evidence and interrogations, seems quite plausible, the last (the means of escape) is certainly guesswork, although a number of small planes left Berlin during the last April days. We know that Ritter von Greim, Mueller, von Buelow and a woman, Hanna Reitsch, left this way.

What happened to the burned corpse of Hitler (probably the corpse of the double) is still bizarre. It was buried in the town of Magdeburg, in an unpaved area at 30–32 Klausenstrasse, then the headquarters of the NKVD. It was later disinterred for another investigation. Finally, in 1970 the Russian Secretary General

Brezhnev agreed to again unearth the remains and have them, this time, completely incinerated and the ashes strewn in the Elbe River near the town of Biederitz (see also Bezymenski's article[14] of 1992). This was done in order to eliminate a possible rallying point for Neo-Nazis.[15] However, there was still no news about the alleged second burned corpse, namely that of Eva Braun.

The Hitler suicide story again resurfaced in 1968, when Lev Bezymenski, a Russian journalist and member of the KGB, wrote a book,[16] trying to tell the Russian version of the events in the Bunker as a counterpoint to the Trevor-Roper story. In his book he stated, citing no supporting evidence, that Hitler after dying from poison, was shot by one of his subordinates (this may well have happened to the Hitler double whose corpse had a gun shot wound to the forehead and, by that time, had been burned).[17] He even discussed the cadaver of one of the burned dogs. Bezymenski identified one as "Blondi", Hitler's favorite. However, the dog that was found had a black pelt with a white under-belly, whereas Hitler's Blondi had a brown pelt. Official Spanish police reports of April 1945 list a "large brown wolfhound" as part of the manifest of Hitler's plane, according to Gregory Douglas.[18] It should be noted here, that there were two additional dogs in the bunker kennel. These two dogs, one belonging to one of Hitler's secretary, C, Christian, were poisoned by SS Sergeant Tarnow, as witnessed by Dr. Schenck.

While Bezymenski in his 1968 book[19] stated that Hitler's body was cremated following the autopsy, and that his ashes were scattered in the winds, he retracted himself and in an article[20] written in 1992 stated that Hitler's corpse had not been burned but buried and re-buried several times.[21] This again throws doubt on the whole Russian suicide and funeral story.

Cornelius Ryan, the well-known author, visited Moscow in 1963 to find the truth about Hitler's fate. He questioned Field Marshall Sokolovsky among other high-ranking officers. The Marshall told him that the Soviet Union considered Hitler dead. Ryan was further told, that the Russians did find a partly burned body near the bunker. A bullet had entered the right temple and blown out some teeth. Despite the missing teeth, Hitler's dentist then identified the remains as those of the Nazi leader.

This again is very curious and quite contradictory. Now, according to Sokolovsky, Hitler was shot in 1945 instead of poisoned as originally stated in the 1945 Russian autopsy, and Hitler's dentist (who was at that time in U.S. custody) made the alleged identification.

An editor for the German magazine *Der Spiegel* also tried to pry more information from the Russians and he was able to interview Colonel Klimenko, the first investigating Russian officer to reach the bunker. However, his statements were confusing and still gave no clarification how the Fuehrer died or what the Russians did with his corpse.

All this implies again the likelihood that the official Russian story of two burned corpses and the alleged autopsy was an elaborate hoax concocted for political reasons.

What shall we make of this? First there is a strong possibility that if a burned body really existed then it could have been another false lead arranged by Mueller's security service, using the body of either the General Burgdorf or Colonel Schaedle, both were said to have committed suicide and neither body was reported to have been found. (SS General Mueller[22] stated to his interrogators that General Burgdorf escaped and that he later worked for U.S. Intelligence, a statement to which his interrogator agreed). Finally, this body could have been the corpse of the bogus Bormann who, according to the Nordon Report, had a badly damaged head.[23] Being the "fake" Bormann would lend credence to the Russian report that this male corpse was poisoned and had a missing cranium. All senior members of the Bunker entourage had been given capsules with cyanide.[24] Remember, Himmler and Goering died this way too. As to the claimed female body, if it really existed, this could have simply been a German women army auxiliary killed in battle, or it could have been the body of a "fake" Eva Braun. Finding corpses in the devastated Berlin of May 1945 certainly posed no problems for the Russians.

It is of interest here to note that SS Major Guensche, a key witness to the alleged cremation, stated to his U.S. CIC interrogators on November 15, 1958, "that I did not see the dead Fuehrer". The valet, Linge, and Chief Pilot Baur, gave similar testimony to U.S. investigators after their return from Russian prisons. Both Linge and Baur asked their Russian jailers whether or not they ever found Hitler's body. They only received evasive answers but were never asked to identify potential corpses.

We already learned that Hitler's chauffeur, Kempka, did not see Hitler's face either. This is pretty strong and independent corroboration that the supposedly burned body was never identified as that of Hitler. A West German television program in November 1971 showed X-ray pictures of Hitler purportedly taken by the German Dr. Giesing in September 1944. On this program, where both Dr. Giesing and the former U.S. prosecutor Robert Kempner appeared, it was shown that these photos differed radically from those described in the Russian autopsy report. A steel pin in the lower right incisor is plainly evident. The false teeth in the lower jaw were not attached to the right incisor. There was also a difference in the number of teeth. Both men also reported that Hitler, indeed, had two testicles. This again implies, that the Russian "two burned body" story was a fake.

Judge Musmano, after hearing about 200 witnesses in 1948, reached the conclusion that Hitler's corpse was never found.[25]

Highly significant is the report that dealt primarily with the disappearance of Martin Bormann from the bunker, but in which vital clues pertaining to Hitler are revealed.[26] This report was written by Captain Otto. N. Nordon, a special assistant to the U.S. Major-General Wm. J. "Wild Bill" Donovan, the head of the OSS, forerunner of the CIA. On May 17, 1945, General Donovan briefly became an assistant to Supreme Court Justice Robert H. Jackson, the U.S. Chief Prosecutor at the Nuremberg War Trials that started in the fall of 1945.[27] Otto N. Nordon had been an attorney in New York before joining the OSS. Donovan thought highly of him and gave him several important assignments in Germany including this report written in Nuremberg and dated November 3, 1945 from which I quote:

Confidential Soviet (Russian) records of their investigative actions reveal the following:
At the time of the capture of the Reich Chancellery during the first week of May, 2, through the 8, special teams of Soviet military and police investigators unearthed two bodies from the garden area.

One body was purported to be that of Adolf Hitler and the other that of Martin Bormann.

The Hitler corpse bore strong physical resemblance to Hitler; was dressed in his uniform and had been shot once in his forehead.

The Bormann corpse had a badly disfigured head making identification impossible. It was dressed in an original uniform of Bormann and had authentic papers in his pockets.

Extensive forensic investigations carried out by Soviet experts at the specific orders of Stalin disclosed that the alleged Hitler corpse was that of a younger, shorter double, while the Bormann corpse was that of a larger man. In this case, the uniform coat was made for a smaller man and did not fit the body.

The Hitler body[28] (then) was partially cremated and then, on orders from Stalin, sent to Moscow. The Bormann body was photographed and fully cremated. A study by our experts of both the Soviet reports and the photographs of the remains conclude that neither corpse was authentic.

The Soviets are now *absolutely convinced* that these bodies were left to provide a *false trail* for investigators. The entire physical area of the Chancellery was probed and excavated by the Soviet specialists *without the discovery of any other bodies or forensic evidence.*

This is very interesting indeed, since it confirms part of the description, regarding the size, given of the Hitler double by SS General Mueller. The Nordon report does not mention later Russian reports that two additional bodies, allegedly of Hitler and Eva Braun, were found on May 5, or six months *prior* to this Nordon report. This too puts the whole "Viking funeral" story for April 30 into question. It also supports the Russian leadership's suspicion that Hitler and Eva Braun escaped and furthermore, it confirms the existence of a conspiracy to arrange Hitler's escape and the murder of Hitler's double. What is additionally intriguing is the attempt to fake the death of Bormann, in order to wipe out his escape trail too. Note the fact that the Germans did not burn the corpse of the bogus Bormann. This too exposes the whole story of the two partly cremated corpses as a farce.

It can, of course, be argued that the information Captain Nordon received was bogus. Yet vital details fit the overall picture. Remember that Captain Nordon did not get this information through official and open Russian channels but from U.S. CIC Intelligence, the British Secret Service (MI5) and from the confidential files of the Soviet Prosecutor for the International Military Tribunal.

Of added significance is the following statement in this report:

> U.S. and British agencies have no specific information about the survival or death of Martin Bormann. The Soviets have had and maintain absolute control over both the former Nazi government centers in Berlin and most key witnesses. *They have so far declined any substantive assistance to outside investigation agencies.* [emphasis added]

Coming back to the Russian investigations in May of 1945, here their whole effort lacked thoroughness. This is a fact also mentioned and criticized by Trevor-Roper. For example, there is no forensic evidence given of the bunker scene. No bullets or bullet fragments were ever found. Stories were later circulated that bloodstains were found on the sofa in Hitler's study, alas of the wrong blood type. However, some evidence was certainly destroyed since according to the testimony of the German bunker Chief Bunker Technician Hentschel, Hitler's anteroom was set afire by SS Captain Schwaegermann, using gasoline. This was done on the orders of SS General Mohnke. This order was carried out during the night from May 1 to May 2. As to the effect of the fire, Hentschel testified that the metal door to the study was red hot and that the rubber air seals were melting. Later that morning after the anteroom was burned, the first Russians entered the bunker. They were female medical personnel who began immediately to loot the place and later left, brandishing some underwear taken from Eva Braun's bedroom. Later on some officers arrived and apparently discovered the significance of the bunker. These events happened hours before Colonel Klimenko and his investigative team arrived. A lot of evidence may have been tainted or destroyed by this time.

An intriguing question is, why did General Mohnke order Hitler's anteroom, to be burned? The logical answer is, to destroy evidence, perhaps of the murder of Hitler's double.

We may conclude from all of this that there was *no positive identification* by the Russians of this partly incinerated corpse (if it really ever existed) that was supposed to have been the real Adolf Hitler.

Perhaps the clearest and most truthful statement is the one given by the Russian Marshall Zhukov after he visited the Reich chancellery on May 3, 1945 and quoted here from Joachimsthaler's book:[29]

> After the chancellery had been taken …we wanted to appraise ourselves on the spot of Hitler's, Goebbels', and other prominent Nazis' suicides. This was difficult, however. No one knew exactly at which place and who had been involved. The statements contradicted each other. Prisoners, mostly wounded, were unsure to say anything about Hitler and his entourage. …We searched in vain for the pyres where Hitler's and Goebbels[30] bodies had been burned. … The way the matter stood, I instantly had my doubts that Hitler had committed suicide.

While the Russians did not co-operate in the investigation by the Western Allies into Hitler's death, and did not accept the Trevor-Roper report, they did allow, on December 3, 1945, the Western Allies to dig in the Chancellery garden. This was done a week later and eight German laborers excavated the grounds. However, all that was found were two hats, one undergarment with Eva Braun's initials and some documents written by Minister J. Goebbels. When the search team tried to return on the next day, the Russian guards refused entry.

Chief Pilot Baur was beaten mercilessly in Russian prisons for refusing to admit that he flew Hitler out of Berlin and to Spain, which, of course, he did not. This again indicates that the Russian Secret Service, then the best in the world, may have had a pretty strong idea of what really happened in Berlin. This comes also across from *Moscow News* of June 2, 1945 saying "Franco Spain War Criminals' Hideout" and from the Moscow report of the *Los Angeles Times* on April 27, 1945 (note this early date) "Nazis flying to Spanish Island". The Russians undoubtedly received word from their agents in Spain about the landing of the German plane on April 27. However without concrete proof, Stalin could not call for an extradition of Hitler. Franco simply would have denied the whole thing. Another reason why the Russian interrogators did not get the whole story about Hitler's flight from the captured bunker crew was according to SS General Mueller, none of the remaining co-conspirators, such as Linge or Rattenhuber, were told of Hitler's final destination after his escape from the bunker.[31] They knew only about his "double" and were sworn to pretend that the real Hitler, together with Eva Braun committed suicide. This is in line with the "need to know" policy, a standing order issued by Hitler. Besides, this way witnesses could not reveal anything under torture.

Later on, and only after Stalin's death and under Krushchev's rule in Moscow, did the Russian Leadership's attitude towards Hitler change. They now conceded that Hitler killed himself and, in 1960, the Kazakhstan *Pravda* published three photographs, one of them supposedly of the dead Hitler. They quoted I. Y. Sianov, a former member of the investigative team that entered the bunker, as saying: "This is a picture of Hitler's corpse, the hysterical maniac had shot himself at the very last moment. I saw the body[32] It lay there with a hole in his forehead. His servants had no time to burn the body as he had ordered them to do."

This then is the first Russian denial of their original (1945) story that they found Hitler's corpse and that it had been burned. The published (1960) photo, of course, was that of the unburned corpse of Hitler's double. However, the Russian official policy was, from then on: "Hitler did not escape, he committed suicide" which, after this date was repeated in numerous Russian publications and encyclopedias. It is now an established part of Russian history. With Stalin dead, the Russian generals finally got their way after all!

This "turn about" of official Russia (by accepting the un-burned corpse of the Hitler double as the real thing) finally put the story of the alleged recovery of two burned corpses to rest and, with it, the so called "Viking Funeral" on April 30, 1945 outside the Fuehrer bunker. This was an event which never happened.

Yet, with all this hoopla, there was still no mention of Eva Braun, or of her corpse.

In order to put speculation to an end, a German court in

Berchtesgaden issued an official death certificate on Hitler on October 25, 1956, concluding "that Hitler took his own life on April 30, 1945 at 3.30 P.M. in the Fuehrerbunker of the Reich Chancellery in Berlin, by shooting himself in the right temple." Without any evidence, the court then stated that his body, together with the uninjured Eva Braun, were found sitting and lying together on a sofa, by Goebbels, Bormann and others (Goebbels and Bormann, being conveniently dead would have been unable to make such a statement). It appears that this court adopted the same legal interpretation, later used by U.S. courts in similar cases, namely: If a story has been repeated often enough, it then becomes a historical fact.

Notes

[1] Bezymenski, Lev, *The Death Of Adolf Hitler*, Michael Joseph, London, 1968

[2] Douglas, Gregory, *Gestapo Chief, The 1948 Interrogation of Heinrich Mueller*, James Bender Publishing, 1995

[3] Douglas, Gregory, *Gestapo Chief, The 1948 Interrogation of Heinrich Mueller*, James Bender Publishing, 1995

[4] Brown, Anthony Cave, *The Last Hero, Wild Bill Donovan*, Vintage Books, a division of Random House, 1984

[5] Douglas, Gregory, *Gestapo Chief, The 1948 Interrogation of Heinrich Mueller*, James Bender Publishing, 1995

[6] Other witnesses claim that the dental records of Hitler were in the airplane that crashed on the way to Bavaria.

[7] Brown, Anthony Cave, *The Last Hero, Wild Bill Donovan*, Vintage Books, a division of Random House, 1984

[8] Brown, Anthony Cave, *The Last Hero, Wild Bill Donovan*, Vintage Books, a division of Random House, 1984

[9] Beschloss, Michael, *The Conquerors*, Simon & Schuster, 2002

[10] Douglas, Gregory, *Gestapo Chief, The 1948 Interrogation of Heinrich Mueller*, James Bender Publishing, 1995

[11] Beschloss, Michael, *Dividing The Spoils*, Simon & Schuster, Inc., 2000

[12] Kilzer, Louis, *Hitler's Traitors*, Presido Press, Inc. 2000.

[13] Joachimsthaler, Anton, *The Last Days Of Hitler*, Cassell & Co., London, 1995

[14] *Der Spiegel* a German Newsmagazine, No. 14, 1992, p. 110

[15] Beschloss, Michael, *The Conquerors*, Simon & Schuster, 2002

[16] Bezymenski, Lev, *The Death Of Adolf Hitler*, Michael Joseph, London, 1968

[17] The discovery of poison also agrees with SS General Mueller's testimony that Hitler's double was first poisoned and then shot.

[18] Douglas, Gregory, *Gestapo Chief, The 1948 Interrogation of Heinrich Mueller*, James Bender Publishing, 1995

[19] Bezymenski, Lev, *The Death Of Adolf Hitler*, Michael Joseph, London, 1968

[20] Schellenberger, Walter, *The Labyrinth*, Memoirs, Da Capo Press, 2000

[21] *Der Spiegel* a German Newsmagazine, No. 14, 1992, p. 110

[22] Douglas, Gregory, *Gestapo Chief, The 1948 Interrogation of Heinrich Mueller*, James Bender Publishing, 1995

[23] Brown, Anthony Cave, *The Last Hero, Wild Bill Donovan*, Vintage Books, a division of Random House, 1984

[24] Actually Prussic Acid.

[25] Musmano, Michael, *Ten Days To Die*, second edition, McFadden Books, New York, 1962

[26] Douglas, Gregory, *Gestapo Chief, The 1948 Interrogation of Heinrich Mueller*, James Bender Publishing, 1995

[27] Brown, Anthony Cave, *The Last Hero, Wild Bill Donovan*, Vintage Books, a division of Random House, 1984

[28] The body of Hitler's double.

[29] Joachimsthaler, Anton, *The Last Days Of Hitler*, Cassell & Co., London, 1995

[30] The corpses of the Goebbels had already been removed by Colonel Klimenko on the prior day.

[31] Douglas, Gregory, *Gestapo Chief, The 1948 Interrogation of Heinrich Mueller*, James Bender Publishing, 1995.

[32] The location of the bullet hole in the forehead almost certainly excludes suicide. The double most likely was drugged or poisoned and then got the *coup de grâce* by means of a pistol shot administered by a member of Hitler's security service.

THE TROUBLE WITH THE TEETH

Any good forensic investigation of a murder, or suicide scene trying to check the victim's identity, includes: checking for fingerprints, checking the victim's blood type, and comparing the dental structures with what is known from previous dental records. Looking at the autopsy report of the alleged Hitler corpse and reported in Bezymenski's book[1] there could have been no fingerprints since the body was supposedly burned. Nevertheless, one might still be able to extract some bodily fluids in order to establish the blood type. This apparently was not done. Checking for the DNA was of course unknown at that time. This then leaves only the victims teeth for identification.

Rather than taking the teeth from the supposed Hitler corpse, according to the testimony of the German dental technician K. Heusemann, the Russians removed bridges from the remains of about 13 to 15 corpses scattered around the bunker exit.[2] Where did these corpses come from? Surly not from the bunker exit, but more likely from the stone pond in the chancellery garden, which was used as a temporary morgue by the German field hospital located near the bunker. Then again we read in Bezymenski's book that the Russian Secret Service took a bridge made of yellow metal and consisting of nine teeth out of Hitler's upper jaw and a single lower jaw containing 15 teeth out of Hitler's skull.[3]

We know Hitler's dentist Dr. Blaschke was in U.S. custody and all of Hitler's dental records were destroyed[4]. Luckily, the Russians were able to locate two of Dr. Blaschke's dental technicians. The first one, Kaethe Heusemann was arrested on May 9, 1945 in Berlin. Here she was shown a gold bridge with facets that was removed from an upper jaw. She also was shown a complete lower jawbone with teeth and bridges plus an additional bridge made for a lower jaw and consisting of synthetic resin with a gold crown. She immediately was able to identify the upper bridge and the lower jaw with the dental work as those belonging to Hitler. The way she described it, the lower jawbone contained one larger and one smaller gold bridge. One should note here that Dr. Blaschke, in his testimony to his U.S. interrogators mentioned only one lower bridge![5] Further according to Dr. Blaschke, the upper bridge was fixed. This means it had to be cut from the upper jaw. This is very strange, why was the upper jaw cut but the lower jaw left intact? The reason may be that these parts were recovered from the partially burned corpse of the double.

Even more bizarre is the fact that Frau Heusemann was able to identify the pristine synthetic resin bridge as that belonging to Eva Braun, perhaps not realizing that her corpse was supposed to have been burned! The bridge could not have come from a cremated corpse, since resin, depending on its type, starts to melt between 300 and 500 degree F, much below the temperatures reached by a gasoline fire.

Later on she was shown seven crates partly buried in the earth. These contained the human remains of the Goebbels family including their children. There was also another crate containing the remains of two dogs.

However, no crates of Hitler or Eva Braun's remains were ever shown to her or to any other German prisoner, despite the fact that L. Brezemensky in his book shows a photo of a box containing some dark matter that was supposedly Hitler's remains.[6]

Later on K. Heusemann was asked again to identify the same bridge work, this time contained in a cigar box.[7] This she did. One should note here that her description of Hitler's dental works differs substantially from that given by Hitler's dentist Dr. H. Blaschke to his U.S. captors. This means that one of the two is certainly lying, or was given the wrong evidence. A similar testimony was given to the Russians by a second dental technician Fritz Echtmann who went essentially through the same interrogation procedure, as did K. Heusemann. Unfortunately, he was briefed, by K. Heusemann prior to his arrest. This happened when the Russians released K. Heusemann for a short time after her initial interrogation. Note, that in 1944 F. Echtmann was asked by Dr. Blaschke to produce a "skeleton denture" from X-ray pictures of Hitler's teeth. This he did and one wonders if this bridgework was then implanted into the mouth of Hitler's double,[8] who was first employed as such in the late summer of 1944.[9] This would have been a perfect cover-up, except for the fact that the double still was shorter and that his ears did not match. Such an implant would point to the possibility that the dental items shown to the technicians were taken from the corpse of Hitler's double instead of from the alleged burned corpse of Hitler himself. Since it seems that the X-ray photo of the skull with the bullet hole in the forehead is really that of Hitler's double (see photo), this explains the many inconsistencies in the autopsy report published in 1968 by Bezymenski. But it makes it quite clear that the subject of the autopsy was Hitler's double rather than the real Hitler. It also confirms the finding of the U.S. professor of dental biology Dr. Reidar Sognnaes that the teeth and bridges used by the Russians for identification matched those shown in the X-ray photo. No wonder they matched, since those teeth and bridges came the same skull belonging to the "double".

This could explain the "cat and mouse" game played by the Russian Secret Police with the two dental technicians, including their long incarceration in Russia. That the bridges most likely came from the initially unburned double also explains Hitler's unburned and pristine golden party badge and the iron cross medal, which was shown to K. Heusemann for identification. These items certainly could not have come from a charred corpse.

Of great importance here too is the fact that Fritz Echtmann the second dental technician was repeatedly questioned (for over one year) by the Russians about the disappearance of Eva Braun, despite the fact that the lower resin bridge was identified as belonging to Eva Braun.[10]

Later on in the 1970s, there was also talk of X-ray pictures of Hitler's head, that supposedly were used to identify the teeth and bridgework, but were reported to have "proved too meager" for diagnosis. Copies of some of these films are at the U.S. National Archives; but none of these have the standard German military and medical information on them.[11] In 1972 Dr. Reidar Sognnaes, a professor of dental biology in California, believed that he was now able to positively identify Hitler's dental work using the above films and by comparing them to the illustrations in L. Bezymenski's' book (showing the skull of the double).

However, Sognnaes concluded that the burned female corpse allegedly found was definitely not Eva Braun!

Part of the so-called evidence of Hitler's identification through his dental works came from an X-ray photo that was supposedly taken by Dr. Giesing on September 19, 1944. This film was found in the U.S. archives but as mentioned before showed none of the typical German markings and dates, as previously noted. What is startling is that this photo clearly shows a nice circular hole in the forehead and between both frontal sinuses, slightly to the left of the center of the head. This hole corresponds exactly with the bullet hole in the forehead of Hitler's double. This then is no doubt an X-ray photo taken by the Russians of Hitler's double's head after he was disinterred on May 2, 1945. A copy of this photo might then have been given to U.S. intelligence as a photo of Hitler's head (perhaps before the true identity of the double was discovered). This, of course puts the whole dental identification further into doubt.

Notes

[1] *Der Spiegel* a German Newsmagazine, No. 14, 1992, p. 110

[2] Joachimsthaler, Anton, *The Last Days Of Hitler*, Cassell & Co., London, 1995

[3] Bezymenski, Lev, *The Death Of Adolf Hitler*, Michael Joseph, London, 1968

[4] Douglas, Gregory, *Gestapo Chief, The 1948 Interrogation of Heinrich Mueller*, James Bender Publishing, 1995

[5] Brown, Anthony Cave, *The Last Hero, Wild Bill Donovan*, Vintage Books, a division of Random House, 1984

[6] Bezymenski, Lev, *The Death Of Adolf Hitler*, Michael Joseph, London, 1968

[7] It was reported that she selected Hitler's bridge from many others in that box. This was not true according to her testimony.

[8] Some of us may have seen a movie in which a gangster, trying to dis-appear, had copies of his teeth implanted into a handy corpse, which then was burned in a fake auto accident. The police then promptly identified the corpse as that of the gangster and declared the gangster as dead.

[9] Douglas, Gregory, *Gestapo Chief, The 1948 Interrogation of Heinrich Mueller*, James Bender Publishing, 1995

[10] One can assume that this was a removable bridge that was simply forgotten by Eva Braun in the bunker while she escaped with Hitler.

[11] Douglas, Gregory, *Gestapo Chief, The 1948 Interrogation of Heinrich Mueller*, James Bender Publishing, 1995.

RUSSIAN INTERROGATIONS

While Stalin's public statements, insisting that Hitler escaped, could be dismissed as politically motivated, or as "personal prejudice" as some historians claim,[1] no such motives could apply to the NKVD officers who questioned (and tortured) the German bunker survivors for years in Russian prisons without being burdened by "cold war" considerations.

Of special interest to us is the fact that the major questions of the interrogators always revolve around two topics: first, any information about Hitler's double, and secondly, the facts surrounding Hitler's escape. Such repeated questioning within the secret confines of the notorious Lubianka prison in Moscow, for example, must have been based on more than mere suspicion.

These secret police officers rightfully questioned the official Western suicide story because they had found no corpse of either Hitler or Eva Braun. The Russian Secret Service rightfully concluded that it was not logical to have, within the limited confinements of the bunker and its garden, a double suicide without finding the two corpses (or at least some remains), a fact that is conveniently ignored in the literature. Gasoline, especially in an open fire, does not generate sufficient heat to completely incinerate all human remains. The body of Dr. Goebbels was still recognizable despite the raging fire. In any case the bones would still be present.

One can safely assume that someone in Hitler's entourage, probably the famous Russian mole, informed Stalin of Hitler's escape[2] on April 22, 1945 and his subsequent replacement by a double. Stalin then most likely passed this information on to his Secret Service.

During the long interrogations of the surviving bunker Nazis, interspersed by torture, at least one of the prisoners broke and admitted that all had sworn to pretend that Hitler and Eva Braun committed suicide in order to cover Hitler's tracks.[3] This of course confirmed the suspicion of the Russian Secret Service (and Stalin's belief) that the pair escaped from Berlin. This breakdown and the confession of some of the prisoners, most likely, were the result of Rattenhuber's admission, in a Moscow prison, that Hitler (undoubtedly meaning Hitler's double) was poisoned and then shot. As Guensche later stated: "Rattenhuber got us in to a lot of trouble".

Nevertheless when these prisoners were finally released in the 1950s to West Germany they maintained their cover story. Why was this? Well, by then the West firmly believed in the suicide story promoted by Trevor-Roper and others. Thus the former prisoners such as Linge, Guensche, and others, were encouraged to stick to their story by their British and American Intelligence debriefers. To admit that the Russians may have been right all along would have been politically impossible at a time when the Korean War raged and the Cold War was at its most intense.

This then is the supreme irony of the whole story.

Here is an example of the Russians' concern about Hitler's flight to Spain that Stalin and his secret service already knew about. There are statements by Hitler's chief pilot, Hans Bauer, given in November 1955 after his release from Russian prisons:[4] "During the Winter of 1945/46 and in the Spring of 1946 I was interrogated in Lubianka prison[5] time and time again, mainly by Commissar Dr. Savieliev. During these interrogations, I was accused time and time again of flying Hitler out of Berlin." He was continually beaten on the head since the Russians did not believe his denials. Again he stated: "I always was accused of flying Hitler out of Berlin."

Heinz Linge, Hitler's valet was questioned in the same vein:

After I was captured on the evening of May 1, 1945, I was questioned by various officers and then taken to Moscow in December of 1945. I was held in Lubianka prison and then transferred to the Butyka prison. Here I was questioned for about two and a half weeks, always at night.[6]

The subject of these interrogations was always the question, was Hitler dead or alive? The talk always was about whether he (Hitler) was flown out. There was also constant talk whether a double had been substituted. I was always required to describe my experience in connection with Hitler's suicide in detail. During the interrogations I was always maltreated.[7]

Finally, he said that when he was temporarily transferred back to Berlin in April of 1946,[8] his Russian captors asked him about Hitler's measurements. This is intriguing since we know that the double was about two inches shorter. This last interrogation happened shortly before the Russian Commission issued their concluding report in May of 1946, stating that there was no corpse of Hitler and that the witnesses were lying. Linge then mentioned that he had a cell mate named Ackermann who told him that Hitler and Eva Braun both took poison and that Hitler then was shot (this probably referred to the fate of Hitler's double).

Now we come to the 1955 statements by Hans Hofbeck, another Russian prisoner and former SS guard. Among other things he stated:

The interrogators in Moscow kept harping on the question, who was Hitler's double? Who shot the double? Who brought Hitler out (of Berlin)?

He then said:

I answered the question about the double by saying there had been a porter in the Chancellery who had borne a resemblance to Hitler. This man had facial features that resembled Hitler's and also a similar moustache and a similar hairstyle. However, he was a little shorter. Otherwise I kept insisting that Hitler was dead and that a double had not been shot and burned in his place.

(The last part was certainly true; the Germans did not burn the double).

During later interrogations (probably part of the famous re-enactment of the bunker episode, in April 1946) he shared a cell with Chief Pilot Baur. Here Baur mentioned to him: "A long time ago, a man from Breslau had been presented who looked very much like Hitler. However, Hitler strictly refused."[9]

Otto Guensche, another bunker survivor was captured on

May 2, 1945 and he too was flown to Moscow. Here he was accused of lying when he stated that Hitler killed himself. At one point they even told him that he let himself be captured on purposed in order to mislead the Russians and to create a false trail.[10]

Here is part of the testimony of the bunker technician Hentschel:

> The interrogations took part mainly at night. During these I was badly mistreated several times. During one of the beatings, my right eardrum was ruptured. I was told that Hitler was still alive... The claim was made that someone else was cremated....[11]

It is clear from these questions of the Russian interrogators that they knew that Hitler flew out of Berlin. What they did not know (and which the captured Nazis did not know themselves) was, where did he go and who was the pilot? It is equally apparent that they knew about Hitler's double, except that they lacked details of who he was and where he was from. It would certainly be illogical to ask these questions if you already had an identified Hitler corpse, as it was claimed later.

One must admit, that it was is a great sign of willpower and loyalty to Hitler, their former leader, that most of the imprisoned Nazis stuck to their agreed suicide story despite the tortures and deprivations in Russian prisons. That some of them broke under torture and admitted to the fake suicide story (as the Russians reported) is also not surprising.

Notes

[1] Joachimsthaler, Anton, *The Last Days Of Hitler*, Cassell & Co., London, 1995

[2] Martin Bormann was this mole, according to Louis Kilzer – *Der Spiegel* a German Newsmagazine, No. 14, 1992, p. 110

[3] From the summary report of the Russian Army investigation into the disappearance of Hitler dated May 1946. See also "The Viking Funeral" chapter.

[4] Joachimsthaler, Anton, *The Last Days Of Hitler*, Cassell & Co., London, 1995

[5] In Moscow.

[6] Joachimsthaler, Anton, *The Last Days Of Hitler*, Cassell & Co., London, 1995

[7] This is not surprising since the Russians already knew that he was lying about the suicide.

[8] For the re-enactment of the alleged cremation in the bunker garden.

[9] This was probably true prior to the double's first use after July 20 1944. Note, that the doubles domicile (Breslau) matches Mueller's description.

[10] Joachimsthaler, Anton, *The Last Days Of Hitler*, Cassell & Co., London, 1995

[11] Joachimsthaler, Anton, *The Last Days Of Hitler*, Cassell & Co., London, 1995

THE FATE OF EVA BRAUN

In order to fully understand the fate of the real Adolf Hitler during those last April days in 1945, we have to consider what happened to his mistress and later wife.

Eva Anna Paula Braun was born on June 12, 1912 in Munich, attended a convent school where she apparently learned French and English, and completed business college in 1928. She got to know Hitler in October 1929 and was with him after the end of 1930, when she worked for the photographer Heinrich Hoffmann. In 1935, Hitler bought her a small house in Bogenhausen. From then on she was often a guest at the Berghof, Hitler's mountain retreat, but Eva had to stay in the background on all official occasions. Her presence was practically unknown to the German public and she had contacts only with a select number of Hitler's inner circle. Incidentally, Eva Braun seems to have been a devout Catholic, since she went regularly to Sunday mass at the Berchtesgaden church, according to the local priest. Despite her faith, she tried to commit suicide at least once by shooting herself with her father's pistol. This was probably caused by frustration over Hitler's long absences and over her virtual isolation from ordinary people. Yet there must have been love and affection between the two. The wife of the janitor in Hitler's Munich apartment told the press after 1945 that "Hitler and Miss Braun were very much in love" and "you can be sure wherever Hitler is – dead or alive – Eva is at his side."

Much has been made of Hitler's sex life, part of it by tabloid writers in the popular press after World War Two and partly by Allied propaganda during the war. There were women who claimed to have been lovers of Hitler, at varying times.[1] All these stories tend to lack credibility, especially if one considers the motivation for spreading such rumors.

Due to lack of information to the contrary, we have to assume that the relationship between Hitler and Eva Braun was monogamous. Just consider the time he spent at head quarters and his frequent state of exhaustion and frustration, at least during the later war years. Hitler and Braun became lovers sometime in 1932 and stayed intimate when time allowed. Later, during the final war years, Eva Braun asked Hitler's doctor Morell if he could not give Hitler some medication in order to strengthen his libido.[2]

Practically no one in Germany knew of Eva Braun's existence during or before the war. Even the Russian generals had no clue. For example, on May 8, 1945, Marshall Zhukov identified Eva as "Hitler's secretary" and on June 6, 1945 as "a cinema actress".

Towards the end of the war Eva Braun visited Berlin on January 19, 1945. She returned on February 2, to Munich and then went back for the last time on March 7, by train to Berlin where she stayed with Hitler till the end. There is an account by a stenographer, Hentschel, who told the Allies that he saw Hitler for the last time on April 22, and Eva Braun was with him (this is the date of her apparent departure from Berlin).[3]

Little is known about her personality but there seems to have been love between these two quite different characters. It is understandable that she chafed under the strain of always having to stay in the background.

As described in Joachimsthaler's book "Eva Braun was pretty, rather than beautiful, had an attractive figure and liked all kinds of sports. In Eva Braun's company, Adolf Hitler could relax. By and large, she never betrayed his confidence and never attempted to influence him in his personal or political affairs."[4] Hitler's housekeeper Anni Winter stated in 1948 that "Eva Braun was not very intelligent." However, we may take this statement with a grain of salt since Frau Winter was only in charge of Hitler's Munich apartment, while Eva Braun was in charge[5] of the "Berghof", Hitler's mountain retreat. There could have been a touch of rivalry at play. In any case C. Schroeder, Hitler's secretary described Eva as "very energetic and resolute."[6]

In the bunker too, she apparently stayed out of the limelight but had meals together with Hitler and some of his secretaries, at least till April 22. (This is in contrast to the behavior of the person pretending to be Eva Braun after this date as we see later).

Let's now go to the days after April 21. According to O'Donnell, when the secretaries J. Wolf and C. Schroeder had left on April 22nd, it was said that: "Eva Braun and Frau Goebbels arrived."[7] Where did this Eva Braun come from? Officially, she should have never left the Bunker since her known arrival there on April 15.[8] Was this Eva Braun another fake? This is a strong possibility, since according to the testimony of the remaining secretaries, and as reported by O'Donnell, Hitler's double refused to take his meals with what we may call the new "Eva". He also mentioned, "one notes here in passing, that on this evening[9] (22, April) the Fuehrer and his mistress were neither keeping the same hours nor sharing the same bed." I assume from the fact that she arrived with Frau Goebbels, that this Eva Braun could in reality have been a member of the Goebbels' household, or one of Goebbels' secretaries.

Here is another hint: again from O, Donnell's book we read a statement from Dr. Schenck (also cited previously) that during his apparently first visit to the bunker on April 29, he was invited later, around 2.30 A.M. on April 30, to a dinner and drinking party at which he was joined by "three quite good looking young women."

While two were secretaries (Junge and Krueger), he was told that the third, a well-dressed strawberry blonde, was Eva Braun. Till then he had never heard of her. He described her "as the real life of the party-like a Rhineland carnival queen". And, "She did not strike me as particularly intelligent. She was banal."[10]

This seems to be the only close description on record of what the (substitute) mistress looked like during the last bunker days.

What is striking is that Schenck describes her as a "strawberry blond", yet all of Eva Braun's photos in existence, including some amateur small color films, show her as a "brunette".

The description of her behavior and demeanor would also seem uncharacteristic of the real Eva Braun, who by then was thirty-three years old. We must remember that everyone in the bunker, who knew her, were involved in the conspiracy. Why not extend this cover-up to his bride too?

It would have been very easy for the real Eva to slip out of the bunker prior to Hitler's famous last walk in the garden at

8.30 P.M. on the evening of April 22. Remember, the last time she was seen was shortly before 5 P.M. on April 22, when she had a short, five-minute private conversation with Hitler (perhaps he was wishing her a safe trip).

One clue to her fate may lie in a letter that was found in the bunker by Russian troops.

The letter was partly burned but was addressed by Eva Braun to her parents, stating that they may not hear from her for a long time, so please don't worry. This certainly does not read like a suicide note, but rather a farewell from a person about to go on a long voyage and who, for obvious reasons, could not communicate from her "safe haven". It appears now that this safe haven was Franco's Spain or Argentina.

It can only be hoped that the "substitute" Eva, that Schenck described, was not murdered too by the Security Service, like the poor double of Hitler and the unfortunate notary, Wagner.

One should mention here that the Russians again had no positive identification of the female body that they allegedly found. Even Reidar Sognnaes while stating in 1972, that he found "excellent odontological evidence for identifying Hitler" did not believe that the female corpse was that of Eva Braun.[11] If she was really poisoned, as Trevor-Roper claims, then what happened to her body? Since there was no body, her fate can only be explained by assuming that she fled with Hitler. This again supports the assumption that Hitler escaped.

Further support to this thesis is given by Marshall Sokolovsky in a statement to Cornelius Ryan, the famous writer, when he visited Moscow in 1963. He said "that there is doubt, that Eva Braun's body has been found."

Finally, here we have head lines from *The New York Times* dated June 9, 1945, announcing: "Zhukov says Hitler Wed Actress In Berlin, May be alive in Europe". And continuing quoting Marshall Zhukov: "We found no corpse that could be Hitler" and he added Hitler *and his bride* "had a good opportunity to get away from Berlin."[Emphasis added by this author].

In conclusion one could state that there was no corpse of the real Eva Braun. This means she escaped from Berlin.

Notes

[1] Knopp, G., *Hitler's Women*, Sutton Publishing Ltd., 2003

[2] Knopp, G., *Hitler's Women*, Sutton Publishing Ltd., 2003

[3] McKale, Donald M., *Hitler The Survival Myth*, Cooper Square Press, 1981

[4] Joachimsthaler, Anton, *The Last Days Of Hitler*, Cassell & Co., London, 1995

[5] This was an un-paid position. She was supported in her task by a young couple.

[6] Christa Schroeder, *Er War Mein Chef*, second edition, Georg Mueller Verlag, Germany, 1985

[7] O'Donnell, James, *The Bunker*, Da Capo Press, 1978

[8] Eva Braun stayed at her apartment at the Chancellery between March 7 and April 15, 1945.

[9] This statement refers to Albert Speer, who after midnight, still tried to see Hitler before his departure around 3 A.M. on 4–23–45. He could not, since Hitler "fell asleep". The sleeping Fuehrer undoubtedly was the double, "Hitler". The real couple had left Berlin 5 hours earlier.

[10] O'Donnell, James, *The Bunker*, Da Capo Press, 1978

[11] McKale, Donald M., *Hitler The Survival Myth*, Cooper Square Press, 1981

WHAT HAPPENED TO GENERAL FEGELEIN?

Here is the story of another actor in the drama playing inside the bunker and yet another mystery.

Hermann Fegelein was born on October 10, 1906 in Ansbach and for a while was with the State Police in Munich. He joined the SS on October 4, 1933 and was commander of an SS riding school in 1937. During the war he advanced to be commander of an SS division till the end of 1943. From January 1, 1944 he was Liaison Officer of the Waffen SS to Hitler. On June 6, 1944 he married the sister of Eva Braun.

According to official versions, Fegelein left the bunker on April 25, 1945, was arrested on April 27, and executed during the night of April 28, 1945. Yet General Mueller[1] stated univocally to his U.S. interrogator that Fegelein flew with Hitler to Spain on April 26, 1945.

What shall we make of this? For more of the story I have to rely on O'Donnell's book,[2] and primarily on Hans Baur's testimony. While his chapter is entitled "The Lady Vanishes" and deals primarily with an apparent lady spy (supposedly the wife of an Hungarian diplomat) who Fegelein was seeing, O'Donnell also describes what happened to Fegelein. According to this story, Fegelein, on April 25, drove out of Berlin, supposedly visiting Himmler at his headquarters in Hohenlychen. He then was supposed to have returned to the bunker on April 26, and then went into hiding. Here comes the bizarre part. Hitler then was supposed to need him on the 27, and at 5 P.M. General Rattenhuber sent a posse after him. They returned empty handed. At 11 P.M. they sent another team under Colonel Hoegel. This time they apparently returned with Fegelein after encountering the lady spy. The prisoner then was handed over to our friend SS General Mueller, who promptly disappeared with him. During the night of April 27, he was supposedly brought back to the bunker and court-martialed by a tribunal consisting of Generals Burgdorf, Krebs and Mohnke. The sentence was death. Finally, Fegelein was supposed to have been executed on April 28, 1945. Joachimsthaler told an almost identical story, citing General Rattenhuber as his main source. In this book Fegelein was supposed to have driven to Fuerstenberg on April 25, to visit another SS Officer, Hans Juettner. The next day on April 26, he then called several times to inquire about the military situation. Joachimsthaler surmises "that this proves he was still in Berlin."[3] As a matter of fact, he could have called from anywhere.

This is a strange story. First of all it is unlikely that Fegelein went to Hohenlychen to see Himmler on April 25. According to Walter Schellenberg, the head of Foreign Intelligence, Himmler spent April 23 and April 24 in Luebeck discussing peace terms with the Swedish Count Bernadotte.[4] Besides, Hohenlychen was about to be captured by Russian troops. Himmler's last headquarters in Wustrow (about one hour southwest of Hohenlychen) was abandoned already on April 22. The last time that Schellenberg talked to Fegelein was on April 21, by telephone, while Schellenberg was visiting Himmler. It seems more likely that he went to Fuerstenberg (as Joachimsthaler stated) to say goodbye to his friend and then drove south.

Concerning the court-martial, General Mohnke (the sole survivor of this episode) later hotly denied to O'Donnell that it ever took place.[5] Other participants died shortly thereafter. Finally, there are no witnesses to the execution and as far as is known, the corpse of Fegelein was never found. We also know from previous testimony that Baur was not a reliable witness and simply repeated hearsay obtained from Rattenhuber.

In his book David Irving gives a much shorter version of what happened to Fegelein.[6] He states "Hitler had hardly seen SS General Fegelein since the previous week. But on April 28, his staff began receiving erratic calls from Fegelein." At about 11.30 P.M. Fegelein was brought back into the bunker in civilian clothes and the Fuehrer ordered him summarily court-martialed and executed.

Those dates do not agree with the previous story.

Yet we still have another version. According to Fest,[7] Fegelein was picked up in civilian dress on April 27, while within the bunker new laments at steadily spreading treachery were heard. As a result "Hitler" had Fegelein subjected to a short, sharp interrogation, then shot in the chancellery by members of his bodyguard. This was supposed to have happened after 10 P.M. on April 28, 1945.

It seems that there is a strong possibility that Fegelein too did escape. He most likely drove to Austria, instead of to Hohenlychen, on April 25,[8] in order to join Hitler, who then departed on April 26, 1945, for Spain[9] according to Mueller. His April 26 phone calls to the bunker, inquiring about the military situation, could have been placed from Hoerching airfield and probably on request of Hitler, who, for obvious reasons could not call himself. The whole story of Fegelein's arrest, court-martial and executions came from statements by General Rattenhuber, one of the main conspirators in Hitler's disappearances. This was no doubt a cover story to explain away Fegelein's flight out of Berlin. That Rattenhuber was not truthful can be seen in the fact that General Mohnke (who was supposed to be a participant) vehemently denied that there ever was a Fegelein court-martial. Colonel Hoegel who was supposed to have arrested Fegelein on April 28, died trying to escape from the bunker and therefore could not testify.

There is another intriguing detail in Gregory Douglas' book. It cites a U.S. CIC report of September 1945 that one Walter Hirschfeld, a CIC agent, was in close contact with Hans Fegelein, Hermann's father, stating that his son Hermann Fegelein was in contact with him and that he was told: "the Fuehrer and I are safe and well."[10]

Notes

[1] Douglas, Gregory, *Gestapo Chief, The 1948 Interrogation of Heinrich Mueller*, James Bender Publishing, 1995

[2] O'Donnell, James, *The Bunker*, DA CAPO Press, 1978

[3] Joachimsthaler, Anton, *The Last Days Of Hitler*, Cassell & Co., London, 1995

[4] Schellenberger, Walter, *The Labyrinth*, Memoirs, Da Capo Press, 2000

[5] O'Donnell, James, *The Bunker*, DA CAPO Press, 1978

[6] Irving, David, *Hitler's War*, Avon Books, a division of Hearst Corp., 1990

[7] Fest, Joachim C., *Hitler*, A Harvest Book. Harcourt, Inc., 1973

[8] He could have gotten out of Berlin by car towards the west. Berlin was encircled only that afternoon.

[9] This may have been the reason for Hitler's delayed departure from Hoerching airfield.

[10] Douglas, Gregory, *Gestapo Chief*, The 1948 Interrogation of Heinrich Mueller, James Bender Publishing, 1995

THE EVIL GENIUS BEHIND IT ALL

SS General Heinrich Mueller, Chief of Hitler's Secret State Police, the GESTAPO, born on April 28, 1900, was the son of a minor official. He completed his primary education and learned to be an aircraft mechanic. In June of 1917, he joined the German Army and in 1918 was assigned to flight training. Mueller then served on the Western front during World War One and earned the Iron Cross and a Bavarian Medal for bravery. After the war Mueller joined the Munich police in 1919. In 1934 he joined the Gestapo in Berlin where he rose rapidly in rank till he was promoted to Lieutenant-General of the Police on November 9, 1941.

He then was head of the German Secret State Police (GESTAPO), in charge of internal security, anti-terrorism and anti-espionage. As such he had a vast network of agents and informers in Germany and also in foreign countries. On top of this he had extensive facilities to listen in on telephone conversations and to open mail.

He also had a mistress, Anna Schmidt, who saw him last in Berlin on April 24, 1945, when he said goodbye to her and gave her some poison in case she wanted to kill herself.[1]

Reading about the last days in the bunker, we see Mueller as the man in the shadows. While his name pops up now and then in the narratives, he never was reported to have talked to "Hitler" after April 22, nor was he ever involved in any official discussions on matters of state during those days. Yet, if we believe his story,[2] he may have been the most important actor on the scene. Here he was busy ferreting the real Hitler out of the bunker and installing a double in his place. He may also have made arrangements for a substitute Eva Braun. In order to reinforce the appearance that the real Hitler and Eva Braun were on the scene, he may have staged a mock wedding ceremony behind closed doors witnessed only by Goebbels and Bormann (both conveniently dead) and attended by a minor magistrate who Mueller then had liquidated for his trouble half an hour later.

Then (probably on April 30,) he had Hitler's double first poisoned and then shot through the forehead.

Finally, even though he did not admit to it, there is a possibility that he also had the substitute "strawberry-blond" Eva Braun done away with.

As we can see from the above activities, Mueller was quite a busy beaver. On top of all of this, he planned his own survival quite early and carefully. He had access to plenty of foreign currency, bogus British pound notes, and false passports. Besides, there were a number of safe houses in foreign countries, among them Switzerland which was most attractive, since it was within easy reach and also a neutral country.

So when everything was arranged, he walked out of the bunker on April 29, 1945 wearing the uniform of a German Air Force Major. That evening he flew out of Berlin using a street in the Tiergarten as a runway. He used an army spotter plane (Fieseler Storch, F 156) with an extra fuel tank and capable of taking off within 50 meters, piloted by one of his agents who had made previous secret landings in Switzerland. Their otherwise uneventful flight passed over Chemnitz and Salzburg and finally landed close to the Swiss border at around 4 A.M. on Sunday, April 30, 1945.

Using a hidden motorcycle with sidecar, the pair went off via walking paths across the Swiss border having changed into civilian clothes.

I have quoted here extensively from Gregory Douglas' book[3] containing the transcript of interrogations of Mueller by American Intelligence Agents that took place in Switzerland between September and October 1948. These transcripts bear the designation MUB–75–96.

We have to question whether these documents are genuine, but judging by other entries dealing with a wide variety of historical happenings during World War Two which Mueller discussed and whose veracity have since been confirmed from other sources (the Hitler disappearance being only a small part of the book), I tend to believe that they are.[4]

As such they constitute new material of very high historical significance.

As to Mueller's personality, he comes across the pages as a something less than pleasant person. His rather macabre humor, his cynicism and his sometimes insufferable arrogance are testimony thereof. He seemed to have been a workaholic and could not stand idleness. This is probably the reason why in September of 1948 he contacted U.S. Intelligence (CIC) to seek employment despite the fact the he was living in a comfortable villa in Switzerland with several servants at his disposal. In his private life he was quite unhappy. One of his two children had Mongoloid features that caused him to avoid social contacts with his peers.

Yet, he must have had a brilliant mind and he certainly was a careful planner and organizer. According to interrogation notes of Ernst Kaltenbrunner, his former superior, "Mueller had a remarkable memory and knew every person who had ever crossed his path and all events."

An example of this was the manner in which he arranged his own "official" death in order to eliminate any trails of him. His agents in Berlin made sure that an official death certificate under number 11706/45 was issued in his name, listing him as "killed in street fighting in Berlin in April of 1945". They then arranged for a grave complete with grave marker, stating: "Our loving father, Heinrich Mueller, born April 28, 1900" at the Berlin-Neukoelln municipal cemetery. This ruse lasted till the 1960s when rumors began circulating that he was still alive.

His grave was formally opened in September 25, 1963 and the remains of three unknown soldiers were found, but no corpse of Heinrich Mueller. The exhumation was requested by the West German Nazi prosecution center in Ludwigsburg.[5] This center learned that Mueller was not dead but was employed by a foreign government. However, they never found out which government it was. Finally, in 1973 the German Government issued a formal warrant for his arrest, which is still outstanding.

Another attempt at disinformation, probably in support of his alleged death, was the return in 1958 of Mueller's papers, effects and decorations to Mueller's family in Munich by the

Information Center of the German Army (WASt). Nobody by then bothered to check the authenticity of these items.

While the German authorities had good reasons to doubt his demise, any efforts to obtain information on Mueller from U.S. Government sources turned out fruitless, despite the appearance that he worked for U.S. intelligence agencies after 1948. German police watched Mueller's family and former secretary closely after 1961. But this too provided no clues as to his whereabouts.[6] There even was an attempted burglary at the home of the Mueller family by two Israeli agents during that time.

Continued pressure by concerned Nazi-hunters during the late 1990s forced President Clinton to order the release of all classified information on former or suspected Nazi War Criminals.

This then lead to a formal opening and dissemination of information on Heinrich Mueller by the National Archives and Records Administration on December 15, 1999. This was done under the auspices of the Nazi War Criminal Records Interagency Working Group established by executive Order 13110 of January 11, 1999.

The released information proved to be very meager. Here is the copy of the summary:[7]

> The file relating to Heinrich Mueller, containing some 135 pages, covers the period from 1945 to 1963. It also contains copies of Nazi produced documents that pre-date 1945. During World War II, Mueller was the head of the Gestapo and the leading administrator in mass killing operations during the period of late 1942 to late 1944. In the files, there were reports, rumors, and allegations that he was working for the Czech, Argentine, Russian and Cuban governments. Rumors are also noted in the files that he was killed in the last days of the war or that he killed himself and his family in 1946.

Though bare of any information within this report[8] that one might consider to be classified, the CIA made one disclaimer, stating: "The Central Intelligence Agency and its predecessors did not know Mueller's whereabouts at any point after the war". This may well be true, as far as it goes. It is entirely possible that

Mueller's contacts in Switzerland were with U.S. Army Intelligence instead of the OSS (the forerunner of the CIA). However, the report does contain two interesting bits of information. One is, that Adolf Eichmann, on trial in Jerusalem (after being spirited out of Argentina), stated that Mueller survived the war. The second one is a statement that "files from the RSHA (Mueller's office) central files vanished in 1945." This certainly confirms Mueller's statement to his U.S. interrogator in 1948[9] that he was in possession of all of his secret files (which Mueller had located in a secret hiding place). He then used these files and the desire on the side of the U.S. to have his material relating to Russian spy networks, as a bargaining tool to avoid prosecution as a war criminal.[10] It should be remembered that this all happened when the "Cold War" was heating up.

Prior to 1948 there was an active effort to find and prosecute Mueller. He was put on a target list on May 27, 1945 by the (OSS) Counter Intelligence War Room. In the monthly summary report of late July 1945 it was thought that his fate was still unknown and it was thought that Mueller remained in Berlin till the remains of his old Nazi staff got reorganized at Hof near Munich. In 1947 British and American authorities twice sought out Mueller's former mistress Anna Schmidt for clues, but found nothing.

The U.S. Intelligence assumed thereafter that Mueller was dead. That is, there were no more formal investigations after 1947 concerning Mueller. This timeframe conveniently coincides with Mueller's first contacts with U.S. agents in September 1948!

Otherwise the released archival information deals only with the usual rumors of Mueller's sightings, rumors of his death, his supposed involvement with Russian intelligence and other less important data. For example, the CIA in 1961 interviewed a defector named Goleniewski, the Deputy Chief of Polish Military Counter Intelligence, who stated that his Soviet supervisors told him that Mueller was picked up by them some time after 1950 and taken to Moscow. This coincides with rumors that Mueller was working for Russian Intelligence. It appears that this rumor was released to throw interested parties off the trail. What possible use would the Russians have had with a man who arrested or killed a good many of their spies, and what use would possession of his files have been on the Russian espionage network? On the other hand, with the exception of the French Resistance movement, his information on British and American spies was relatively meager. He would have been of little use to the Russians and if caught, he certainly would have been executed.

On the other hand, a man with knowledge not only of European Communist spy networks, but also of Communist activities within the U.S.A., would have been of extremely high value to not only the U.S. but also to British Intelligence services. I let the reader decide on which side Mueller did wind up.

The CIA report finally concluded: "that Mueller likely died in early May1945." The reader may note, that this is the opinion of the three authors of the report and not necessarily that of the CIA.

Any admission that Mueller, a wanted war criminal, might have worked for the U.S. Government would certainly be extremely embarrassing. It is for this reason that any information contrary to the above "formal" statements will probably never see the light of day.

We can safely assume that Mueller by now has died of natural causes, since he was born in the year nineteen hundred. Whatever secrets he may still have had, he probably took them with him to his grave.

Notes

1. Record Group 263: Records of the Central Intelligence Agency, Records of the Directorate of Operations. Analyses of the Name File of Heinrich Mueller

2. Douglas, Gregory, *Gestapo Chief, The 1948 Interrogation of Heinrich Mueller*, James Bender Publishing, 1995

3. Douglas, Gregory, *Gestapo Chief, The 1948 Interrogation of Heinrich Mueller*, James Bender Publishing, 1995

4. See Appendix.

5. Record Group 263: Records of the Central Intelligence Agency, Records of the Directorate of Operations. Analyses of the Name File of Heinrich Mueller

6. Record Group 263: Records of the Central Intelligence Agency, Records of the Directorate of Operations. Analyses of the Name File of Heinrich Mueller

7. Record Group 263: Records of the Central Intelligence Agency, Records of the Directorate of Operations. Analyses of the Name File of Heinrich Mueller

[8] Record Group 263: Records of the Central Intelligence Agency, Records of the Directorate of Operations. Analyses of the Name File of Heinrich Mueller

[9] Douglas, Gregory, *Gestapo Chief, The 1948 Interrogation of Heinrich Mueller*, James Bender Publishing, 1995

[10] One should note here that Gregory Douglas' book about the Mueller interrogations was written already in 1995, or four years before the U.S. Archive release stating this important information.

WHAT SHALL WE MAKE OF ALL THIS?

While the reader may decide what is true or false in this story, here are my own thoughts of what really happened between April 20, and May 2, 1945.

My conclusion is that the "historical truth" about Hitler's death is based on an elaborate hoax concocted by some of Hitler's close associates and members of his security services and by an inadvertent cover-up, and the ineptitude of the U.S. and British Security Services and their unwillingness to believe Stalin. What seems to be true is that both Hitler and Eva Braun (and probably Hermann Fegelein) escaped from Berlin and reached Spain by plane.

While the previous evidence leaves a lot of questions and is full of contradictions, we now have Mueller's statements[1] that tend to fill in neatly many of the holes in this story and which provide us with convincing arguments that Hitler and his wife left Berlin on April 22, 1945.

We know that the Nordon Report, the report of the Russian investigation commission (May 1946) and Mueller's interrogation report, all deny that two partly-cremated bodies ever existed. It is therefore quite possible that this was evidence invented by the Russian Intelligence service (NKVD) later, in order to cover up the fact that Hitler and Eva Braun escaped from under their noses.

The Russian prisoners (Baur, Linge, Guensche and Rattenhuber) having found out that the Russians discovered that Hitler's double was a fake, later corroborated this story, once they were released from Russian prisons. It was convenient that this suicide tale also coincided with their own agreed cover story to explain the disappearance of Hitler. One may ask at this point, why did the Germans not cremate the corpse of Hitler's double?

The idea of course was to make the world believe that Hitler was dead indeed. However, there needed to be a back-up plan in case the Russians should discover the true identity of the double (which they did). This then was the story of the double suicide, with the corpses burned to ashes, explaining why no remains were found. One should also remember that all these witnesses were afraid of later prosecution for the murder of Hitler's double. The suicide story, once accepted, would tend to eliminate this possibility.

Note, these prisoners were only released after 1953, that is, after the death of the Russian Dictator, Joseph Stalin. Stalin himself was convinced that Hitler escaped Berlin even though he had no absolute proof. Releasing the prisoners with the suicide story while Stalin was still alive would have been politically "incorrect" to say the least. The same reason applies to the book by Lev Bezymenski[2] supporting a suicide story albeit in a different version than the one by Trevor-Roper. It was only published in 1968.

We have to realize that a very interesting power struggle went on between the victorious Russian Generals who wanted closure in the Hitler affair and who were understandably embarrassed by his escape, and their boss Marshal Stalin who apparently knew better. This desire for closure led the Russian generals early on to announce that they had found Hitler's body (even if it turned out to be only that of his double). But they were rebuked by Stalin, who ordered them to admit that Hitler had indeed disappeared.

This struggle was only resolved in the generals' favor well after Stalin's death in 1953 through the release of the clumsy, and widely discredited, autopsy report shown in Bezymenski's book. While the Russian military never could show conclusive proof that Hitler committed suicide in the bunker, they got at least full support from the Western press and from a multitude of history writers. The honor of the Russian generals was restored, at least in the eyes of Russian history books.

Coming back to the German prisoners; first, they would have gladly supported the Russian suicide story since it would have fitted nicely with their own attempts to cover up Hitler's escape. The time elapsed between their Russian interrogation and their final release from prison (after which they then made their statements to U.S. and British interrogators) can explain some but not all of the many contradictions in the individual stories of each prisoner regarding the suicide and the "Viking Funeral". As to Kempka,[3] the only important eyewitness who was available to the Western Powers in 1945, we know that he told his U.S. and British interrogators anything "they wanted to hear". And they certainly wanted to hear that Hitler committed suicide!

Such a story was the perfect way to contradict the apparent fact that Hitler escaped from Berlin. There was only one problem, no identifiable body was found. To get around this problem it was then suggested that a corpse was found but burned. Again, why not include a female burned body too, since by now everybody knew that Eva Braun was in the bunker? The second problem was how to tie the burned and unrecognizable corpses to Adolf Hitler and Eva Braun. One solution here was to use the then later partly burned[4] corpse of the Hitler double as a prop in their clumsy autopsy effort. This meant, of course, that there had to be a funeral by the bunker Nazis and an effort to burn the body. This certainly is the origin of the story of the "Viking Funeral" although, as is well known, any false testimony usually breaks down if there are too many witnesses involved. There invariably will be small but important details that will differ from one statement to another, as we have seen in this case, thereby exposing their testimony.

If we consider that the whole story of Hitler's and Eva Braun's suicide, with the subsequent partial burning, and burial of the bodies, was an elaborate fake supported by the Russian Secret Police in order to imply that Hitler died instead of escaping, then we also understand their reluctance to invite, in 1945, independent forensic experts, and to publish exact supporting forensic evidence, such as autopsy photos, X-rays and reports. The reason for not doing this is now clear, there simply were no bodies of Hitler and Braun, and hence there could not have been any exact forensic reports.

We have now four independent reports insinuating that Hitler disappeared from Berlin on or about April 22, 1945:

1. The findings of the Russian NKVD committee charged to investigate Hitler's fate and originally chaired by Marshal Zhukov (and later by Secret service Chief Beria) in a report issued in May 1946;[5]

2. The Nordon report dated November 3, 1945;[6]

3. The statement by W. F. Heimlich,[7] the former Chief Intelligence Officer for the U.S. Army in Berlin;[8]

4. The testimony of SS General Heinrich Mueller.[9]

Of course, to this we have to add the many pronouncements by Stalin that Hitler escaped from Berlin.[10]

There will be an effort calling the Mueller interrogation transcripts a hoax, or the product of some person's imagination, especially when one considers the politically explosive nature of these writings (see Appendix).

However, even conceding this possibility, there is sufficient evidence here to form some basic and sound conclusions. The Mueller statements are, so to speak, only the icing on the cake.

It should be noted here, that while General Rattenhuber and his staff of about fifty remaining bodyguards reported directly to Hitler, while the latter was in the bunker (having sworn a personal loyalty oath to him), this changed after Hitler left on April 22, 1945. Now Lieutenant-General of the Police, Heinrich Mueller, who was also head of the German Secret Police, took charge, being the superior officer of Rattenhuber. Mueller controlled all the strings in the bunker from the evening of April 22, 1945 till the end.

Here than are the salient points confirming Hitler's escape and using nothing but the accepted and printed testimony of what happened, excluding Mueller's statements:

- There was no positive identification of any Hitler corpse in Berlin, at least not one that would stand up to a thorough forensic investigation.

- There was no identified corpse found of Eva Braun.

- There is no dispute that a corpse representing a Hitler double was found. This indicates a conspiracy and likely a murder. The Nazis did not cremate this corpse.

- All surviving witnesses agree to the reduced mental state, erratic behavior, poor physical condition, short attention span, different sleeping pattern and loss of power, of the man remaining in the bunker after April 22, 1945, all very uncharacteristic of the real Hitler. This points to a different person, a double.

- There was no physical evidence found from the room where the supposed double-suicide happened except some bloodstains that could have come from the double or from the suicide of some German officer.

- There may have been a mock funeral of two unidentified corpses and two dogs, likely arranged by Russian investigators to verify the story of their prisoners. However, this does not disprove Hitler's disappearance.

- The testimony of important surviving eyewitnesses conflict which each other on the most important details.

- There was the conviction and the many statements of Stalin, indicating that Hitler did escape.

- The last photograph taken of the Fuehrer was on April 21, 1945, the day prior to his departure from Berlin. There exist no photos thereafter.

- It was confirmed by several witnesses that the person supposed to be Hitler, after April 22, could not make any important decisions on his own (these were made primarily by Goebbels and to a lesser extend by Bormann). In addition, this "Fuehrer" was under virtual house arrest.

- That the assumed Hitler and Eva Braun never shared the bed or lunch table after April 22, contrary to the habit prior to this date.

- According to the Nordon report, and the Russian investigative report of May 1946, no bodies of Hitler and Eva Braun were ever found by the Russians.

- We now have two independent reports that a German type JU 290 airplane did land on April 27, 1945, in Barcelona, Spain.[11;12]

- That the Russians in 1960 published a photo of (the unburned) corpse of Hitler's double, indicating it to be the real Hitler, contradicting their previously announced finding of a burned corpse (see also above).

- That Hitler's will must have been pre-signed, since it appoints Dr. Goebbels as Chancellor, even though it was known that Goebbels wanted to commit suicide, which happened two days after the supposed signing of both wills.

- We know that Colonel Baumbach, who was supposed to be in charge of Hitler's plane to Spain, was in Berlin on April 21, 1945, in order to obtain a civilian flight certificate (likely needed in Spain or Argentina). He did this a day before Hitler's departure from Berlin.

- The Russian high-level investigation commission concluded in May 1946, after a one year investigation and after an exact re-enactment at the scene in April 1946, that the surviving eyewitnesses were lying and that there was no cremation of Hitler and Braun.

To summarize, the whole double suicide and subsequent cremation story was not based on verifiable physical evidence but on the statements of only four surviving ardent Nazis who all were lying, according to their Russian interrogators. The German court that investigated Hitler's death stated that none of the 40 witnesses interrogated ever saw the actual body.

When we add to this the Mueller story that Hitler and Eva Braun flew out of Berlin on April 22, 1945 and that he, Mueller, was the organizer, confirming the above conclusions, then we have an open and shut case ready for any jury.

Notes

[1] Douglas, Gregory, *Gestapo Chief, The 1948 Interrogation of Heinrich Mueller,* James Bender Publishing, 1995

[2] Bezymenski, Lev, *The Death Of Adolf Hitler*, Michael Joseph, London, 1968

[3] Note, all events relating to the period after March 1945 where not written by Kempka but by Erich Kean, and therefore lack all historical significance. 39 Erich Kempka, *Die Lehrter Tage Mit Adolf Hitler*, Verlag K. W. Schuetz K. G. Germany, 1976.

[4] He was partly cremated by the Russian troops the day after he was found intact.

[5] Joachimsthaler, Anton, *The Last Days Of Hitler*, Cassell & Co., London, 1995

[6] Douglas, Gregory, *Gestapo Chief, The 1948 Interrogation of Heinrich Mueller*, James Bender Publishing, 1995

[7] Heimlich wrote in the introduction of a book by Herbert Moore and James W. Barret, entitled: *Who Killed Hitler*, (Booktab Press, New York) the following: "I was assigned to find Adolf Hitler or his body immediately after the entry of U.S. forces into Berlin. I can positively state that I found neither Hitler nor his physical remains. Despite a thorough search in the area, I was unable to discover any proof that his body had been burned, nor was I able to find persons who were eyewitnesses to Hitler's final days in the Chancellery... I can only stress the fact that I was not successful in finding reliable eye witnesses for Hitler's activities after 22 April 1945... nine days before his alleged suicide... My own personal conclusion is that as far as Hitler's fate is concerned, everything after 22 April 1945 is a mystery..."

[8] Joachimsthaler, Anton, *The Last Days Of Hitler*, Cassell & Co., London, 1995

[9] Douglas, Gregory, *Gestapo Chief, The 1948 Interrogation of Heinrich Mueller*, James Bender Publishing, 1995

[10] Michael Beschloss, *Dividing The Spoils*, Simon & Schuster, Inc., 2000

[11] Douglas, Gregory, *Gestapo Chief*, the 1948 Interrogation of Heinrich Mueller, James Bender Publishing, 1995

[12] Hentschel, Philip, *The Nuclear Axis, Germany, Japan and the Atomic Bomb Race*, Sutton Publishing Limited, 2000

CHRONOLOGY OF EVENTS

It might help the reader to follow the trail better by looking at the factual, or the most likely sequence of events while unburdened by too much detail:

FRIDAY, APRIL 20, 1945. Adolf Hitler celebrated his 56, birthday. He reviewed a group of Hitler Youth members who had received decoration for bravery. This was followed by a reception at the new Reich Chancellery involving most of his close Generals and Ministers.

SATURDAY, APRIL 21. Hitler held his last major War situation conference with his generals.

Last photograph was taken of him standing in front of Chancellery.

SUNDAY, APRIL 22. AFTERNOON. Hitler said goodbye to most members of his staff and ordered them, his top generals and all ministers (except Bormann and Goebbels) to leave the same day. About eighty persons departed by ten airplanes to Salzburg, in Austria.

Shortly before 5 P.M. Eva Braun talked to Hitler and then seemed to have disappeared.

At 8.30 P.M. Hitler walked out of the garden behind the bunker and flew out of Berlin by helicopter. All of Hitler's private papers were burned. Hitler also relinquished all of his military commands.

MONDAY, APRIL 23. At 3 A.M. in the morning Albert Speer flew out of Berlin. According to Chief Pilot Baur, there were now only about twenty persons left from the old bunker group.

TUESDAY, APRIL 24. SS General Mueller says goodbye to his mistress in Berlin.

WEDNESDAY, APRIL 25. SS General Fegelein leaves the bunker.

Later that day, Berlin is encircled by Russian troops.

THURSDAY, APRIL 26. In the evening Hitler, Eva Braun and SS. General Fegelein flew to Barcelona, Spain, from Hoerching (near Linz, Austria) airfield.

FRIDAY, APRIL 27. Hitler's plane lands in Barcelona.

SATURDAY, APRIL 28. General Weidling's break-out plan rejected by Goebbels. Hanna Reitsch and Air Force Marshall Greim flew out of Berlin after a short visit.

SUNDAY, APRIL 29. Shortly after midnight on this Sunday morning there was supposed to be a wedding ceremony with Adolf Hitler and Eva Braun behind closed doors. There are no surviving witnesses to confirm this.

General Mueller claimed that the wedding certificate that was found was pre-written and pre-signed by Hitler and Eva Braun. At 4 A.M. the same morning Hitler's political and his personal will were signed by witnesses and perhaps dated (Hitler apparently signed these papers prior to 22, April 1945). At 11 P.M. General Mueller flew out of Berlin to Switzerland via spotter plane.

MONDAY APRIL 30. Final lunch for the Hitler double at 12.30 P.M. At approximately 3 P.M. Youth Leader Max Axmann arrived from his battle station in Berlin trying to see Hitler. He was told it was too late. He was later allowed to see two covered corpses. This could have been the Hitler double and the bogus Martin Bormann. It appears the double had by now been shot and carried out for burial in a shallow grave. The following is of questionable veracity: At about 4 P.M. two additional corpses and those of two dogs were carried into the garden and partly cremated. No human remains were later found.

TUESDAY. MAY 1. During the morning, Goebbels sent a telegram (also signed by Bormann) to Admiral Doenitz[1] (who was in Flensburg) that Hitler died the previous day (no cause of death was stated). At around 8.30 P.M. Dr. and Mrs. Goebbels walked up the stairs to the bunker garden and there committed suicide. Their bodies were burned with gasoline. After 8.45 P.M. most surviving bunker inhabitants tried to break out of encircled Berlin, only a handful made it. At about 9 P.M. SS Captain

Schwaegermann burned Hitler's study, the alleged site of the double suicide, on orders of SS General Mohnke.

WEDNESDAY, MAY 2. The remaining German troops under General Weidling surrender.

In mid-morning the first Russian troops arrived at the bunker. At 2 P.M. Russian investigators arrive and discover the burnt bodies of Dr. and Mrs. Goebbels.

In mid-afternoon Russians discover the body of Hitler's double and that of the bogus Martin Bormann. The "Hitler" corpse is displayed in the Chancellery.

THURSDAY, MAY 3, Russian soldiers are said to have discovered two more semi-cremated bodies; they then were re-buried. (We only have the Russian statements but no corroborating evidence).

SATURDAY, MAY 5, Colonel Klimenko is reported to have the two partly burnt corpses (thought to have been the Hitler couple) disinterred and sent for an autopsy. (There are neither photos nor any other corroborating evidence that these corpses really existed).

Notes

[1] Doenitz, Karl, *Memoirs, Ten Years and Ten Days*, Da Capo Press, Inc., 1997, pp. 440–441

CAST OF ACTORS IN THE BUNKER DRAMA

Here is a listing of the main participants in the drama that was played out between April 23 and May 2, 1945:

ARTHUR AXMANN. Escaped 1st May 1945, Captured by U.S. Army later in 1945. After breaking out with others, he found shelter with a women friend in Berlin.

GENERAL HANS BAUR.★ Wounded while trying to escape May 2, 1945, made Russian prisoner. He too joined the group that was trying to escape Berlin on May 1 but was shot in both legs, his chest and in one arm, during the early morning of May 2. His capture, together with that of SS General Rattenhuber, was announced by the Red Army on May 6. He was then accused by the Russians of flying Hitler to Spain.

COLONEL N. VON BUELOW. Escaped April 29, 1945 to western part of Germany. Arrested by the British Army in the spring of 1946, he was released on May 14, 1948.

MARTIN BORMANN.★ He supposedly died on May 2, 1945. Skeleton found December 8, 1972. Bormann took part in the evening of May 1st breakout from the bunker. Arthur Axmann testified that he saw Bormann's body apparently poisoned near the Lehrter Railway station at about 3.30 A.M. on May 2.

GENERAL W. BURGDORF.★ Assumed to have committed suicide in bunker, probably escaped and worked for U.S. Intelligence after 1945.

GERDA CHRISTIAN. Was raped during breakout attempt on May 1 but ultimately was smuggled out of Berlin by some British soldiers.

DR. JOSEPH GOEBBELS.★ Committed suicide in bunker on May 1, 1945 at about 8.30 P.M. His body together with that of his wife was cremated using gasoline.

MAJOR O. GUENSCHE.★ Captured by Russians May 2, 1945. He was flown to Moscow on May 9, and imprisoned there. He was released in 1956.

PROFESSOR W. HAASE★ Captured by Russians, died in prison in the Fall of 1945.

WALTER HEWEL. Committed suicide May 2, 1945. He too made the breakout in the evening of May 1 but got only as far as the Schultheiss-Patzenhofer brewery. Here he shot himself.

GERTRUD JUNGE. Was raped and injured during the breakout attempt of May 1. She was ultimately rescued by a Russian Major who kept her in Berlin for a year.

ERICH KEMPKA.★ Escaped Berlin, he was hiding in the apartment of an alleged prostitute but was later captured in West Germany by the U.S. Army on June 8, 1945. He was released in October 1947.

GENERAL H. KREBS.★ Committed suicide in the bunker on May 1, 1945. Prior to killing himself, he left the bunker just after midnight on May 1 and went to see Soviet General Chuikov under a flag of truce in order to negotiate. Here he told Chuikov of "Hitler's" suicide and wanted to discuss surrender terms. No agreement was reached and Krebs returned to the bunker in the early afternoon.

HEINZ LINGE.★ Captured by Russians May 2, 1945, made prisoner. Another member of the breakout group of May 1. He was convicted and sentenced to 25 years of hard labor, but was released in 1955.

SS GENERAL MOHNKE.★ Captured and imprisoned by the Russians May 2, 1945. He led the group of escapists on the evening of May 1, but got only as far as the Schultheiss brewery. He surrendered the next day to Russian troops. On May 9, he was flown to Moscow and imprisoned. He was finally released in 1955.

SS. GENERAL H. MUELLER.★ Escaped Berlin by plane for Switzerland, April 29, 1945. Declared himself to be dead. Reportedly worked for U.S. Intelligence after October 1948.

SS-GENERAL RATTENHUBER.★ Wounded trying to escape May 1, 1945. Russian prisoner.

GENERAL REIMANN. Captured and made prisoner of war by Russians May 2, 1945. He died in a Russian prison in 1955.

FRANZ SCHAEDLE.* Chief of Hitler's bodyguard, severely wounded, committed suicide by shooting himself in the bunker.

HERR SILLIP. (Hitler's double). Killed in the bunker probably on April 30, 1945. His body was initially displayed in the Chancellery and later cremated by Russian Intelligence Services. His corpse is the only realistic photo of "Hitler" ever published by the Russians.

COLONEL DR. STUMPFEGGER.* Committed suicide May 2, 1945. He took part in the May 1 breakout. Athur Axmann later reported that he found his body near the Lehrter Railway station at approx. 3.30 A.M. on May 2. He seemed to have taken poison.

GENERAL WEIDLING. Surrendered Berlin May 2, 1945, made prisoner of war.

It is interesting to note that out of the twenty-two most important eyewitnesses ten died (names underlined) before they could testify in the West. One disappeared.

Names with asterisk denote likely co-conspirators in the disappearance of Hitler and the subsequent murder of Hitler's double.

As a historical foot note, and reported by O'Donnell,[1] the Russians staged a filmed re-enactment in 1946 of the Russians idea of the last bunker scenes using the imprisoned Generals Mohnke, H. Baur, J. Rattenhuber, Major Guensche, Dr. Schenck, J. Hentschel and two former SS bodyguards as actors. This occurrence was actually observed by James O'Donnell. No conclusions from these re-enactments were ever reported. However one could conclude that this was a reenactment in order to check out the different stories of the German witnesses in regards to the "Viking Funeral". It did not go too well since shortly there after the Russian investigation committee declared that these witnesses were lying to cover Hitler's escape.

The overall irony in all of this is that after Hitler escaped from Berlin, he tried to tell the world through his loyal followers that he committed suicide and that his corpse was burned to ashes in order to cover his trail. In that he succeeded, thanks in great part to the help of the Western media and governments. While the Russians knew better, they too finally conceded to world opinion, albeit twenty years later.

Notes

[1] O'Donnell, James, *The Bunker*, Da Capo Press, 1978

APPENDIX

As the reader will notice, the testimony of SS General Heinrich Mueller is by far the most revealing and can be considered politically the most explosive information on, not only Adolf Hitler's escape from Berlin, but also other matters.

One can therefore expect major efforts by some authors who staked their reputation on previously established historical assumptions, to declare the whole interrogation report either a hoax, or a forgery.

John Lucas (the author of a book on Churchill) recently made such an effort in an article that appeared in the Nov./Dec. issue of *American Heritage* (AN 7682532) entitled, "The Churchill-Roosevelt Forgeries".

In this article, John Lucas made attempts to prove that the wartime German intercepts of radiotelephone conversations on November 26, 1941 between Roosevelt and Churchill, mentioned in Mueller's interrogation report, were "cleverly falsified and were forgeries."

One of Luca's arguments was that Churchill did not call President Roosevelt "Franklin".

However, according to John Meacham, they always called themselves by their first names, except during official functions and meetings.[1]

The second argument is that Churchill did not employ coarse language, and he cited as witness a dear old lady who acted as censor and who overheard the phone conversations.

While one must admire the loyalty and even more the excellent memories of this lady, the facts are otherwise. In any case, she confirmed that these secret (and never officially published or acknowledged) phone conversations existed. Incidentally, Roosevelt's son Elliott also confirmed this fact.[2] Churchill indeed was known to use salty language and could be quite foul mouthed on occasions.[3]

The next argument of Lucas was that Mueller, while escaping Berlin, had only minimum luggage (and therefore could not have taken all his files along). This is certainly true. However. Mueller arranged to transfer all is files out of his Berlin office (prior to his escape flight) and to a secret hiding place. The story of his missing office files was recently confirmed in a published CIA report.

Lucas then said that Churchill never called American anti-communists "Fascists".

While this is generally true, we have to consider this remark (on November 26, 1941) in context. It referred to Roosevelt's campaign for his third term in office and he was concerned about the vote of the sizable Italian-American population. These certainly were potential fascists instead of Nazis.

Another argument put forward was that when Churchill warned Roosevelt on November 26, 1941 that the Japanese would attack Pearl Harbor, that this could not possibly be true, stating that Churchill would not have said that the Japanese fleet would be sailing to the *east*.

The facts are that the Japanese Fleet did indeed set sail on November 27, and then headed east with the aim of attacking Pearl Harbor in Hawaii. The fleet only turned south and towards Hawaii on December 6, 1941 The actual attack then occurred on December 7.

Note the accuracy of Churchill's warning. When he made his warning on November 26, it was already November 27 in Japan.[4]

Finally, Lucas cites that German Army transcripts of the intercepts did not mention Mussolini by name (referring to a phone conversation in 1943 shortly after Mussolini was arrested in Italy on orders of the King). This is strange since those German intercepts were so secret that only Hitler and perhaps Himmler (Mueller's boss), had a copy. It is likely that this so-called Germany Army copy was a poor translation, or a fake document. In any case the book by G. Douglas[5] makes no reference to any "Army Transcript" and indicates that Mussolini was certainly named in the conversations between Churchill and Roosevelt.

On must remember, that these intercepts (incidentally made with the aid of a U.S. decoding machine sold by AT&T to Germany prior to the war) were so frank and confidential that neither Roosevelt nor Churchill kept a written record of their conversations. Thus we have no way to prove or disprove the correctness of these intercepts. Nevertheless, we know that these phone conversations occurred and that the Germans intercepted them starting on September 1, 1941.[6]

One can conclude that Mr. Lucas' arguments to prove that the intercept were forgeries were ill researched and are spurious.

Notes

[1] Meacham, John, *Franklin and Winston*, Random House, Inc., 2003

[2] Roosevelt, Elliott, *As He Saw It*, Duell, Sloan and Pearce, New York, 1946

[3] Cannadine, David, *In Churchill's Shadow*, Oxford University Press, 2003, pp. 57–58

[4] Weill, Susan, *Pearl Harbor*, Tehabi Books, Inc., 2000

[5] Douglas, Gregory, *Gestapo Chief, The 1948 Interrogation of Heinrich Mueller*, James Bender Publishing, 1995

[6] Irvine, David, *Hitler's War*, the Viking Press, New York, 1977, p. 419

REFERENCES

American Heritage, *Pictorial History Of World War II*, Heritage Publishing Co. Inc., 1966, p. 574

Barnett, Correlli, *The Collapse Of British Power*, Sutton Publishing Ltd., 1997, p. 591

Baumbach, Werner, *Broken Swastika*, Dorset Press, 1992

Beschloss, Michael, *Dividing The Spoils*, Simon & Schuster, Inc., 2000

—— *The Conquerors*, Simon & Schuster, 2002

Bezymenski, Lev, *The Death Of Adolf Hitler*, Michael Joseph, London, 1968

Brendon, Piers, *The Dark Valley*, Alfred A. Knopf, New York, 2000

Brown, Anthony Cave, *The Last Hero, Wild Bill Donovan*, Vintage Books, a division of Random House, 1984

Chronik Der Deutschen, Chronik Verlag, Germany, 1983, p. 926

Coates, Steve, *Helicopters Of The Third Reich*, Ian Allan Publishing, Ltd., 2002

Dix, J. and Calaluce, R., *Forensic Pathology*, CRC Press, LLC, 1998, p. 83

Der Spiegel a German Newsmagazine, No. 14, 1992, p. 110

Doenitz, Karl, *Memoirs Ten Years and Ten Days*, Da Capo Press, Inc., 1997, pp. 440–441

Douglas, Gregory, *Gestapo Chief, The 1948 Interrogation of Heinrich Mueller*, James Bender Publishing, 1995

Fest, Joachim C., *Hitler*, A Harvest Book, Harcourt Inc., 1973

Goni, Uki, *The Real Odessa*, Granta Books, London, 2002

Henshall, Philip, *The Nuclear Axis, Germany, Japan and the Atomic Bomb Race*, Sutton Publishing Limited, 2000

HERMANN HISTORICA, 45th Auction Catalog for October 17–18, 2003, pp. 346–341, "Oberst Baumbach Memorabilia"

Irving, David, *Hitler's War*, Avon Books, a division of Hearst Corp., 1990

Joachimsthaler, Anton, *The Last Days Of Hitler*, Cassell & Co., London, 1995

Kempka, Erich, *Die Letzten Tage Mit Adolf Hitler*, Verlag K. W. Schuetz K. G., Germany, 1976

—— *"Erklaerungen von Herrn Erich Kempka vom 20–6–45 und Ergaenzende Erklaerungen des Herrn Erich Kempka"* vom 4–7–45, given in German to the U.S. investigating officer Harry Palmer

Kershaw, Ian, *Hitler*, W. W. Norton & Co., 2000

Kilzer, Louis, *Hitler's Traitors*, Presido Press, Inc., 2000

Knopp, G., *Hitler's Women*, Sutton Publishing Ltd, 2003

Lucas, John, *The Duel*, Ticknor & Fields, Houghton Mifflin Company, 1991

McKale, Donald M., *Hitler The Survival Myth*, Cooper Square Press, 1981

Musmano, Michael, *Ten Days To Die*, second edition, McFadden Books, New York, 1962

O'Donnell, James, *The Bunker*, Da Capo Press, 1978

Overy, Richard, *Interrogations, The Nazis In Allied Hands 1945*, Penguin Putnam, Inc., 2001

Petrova, Ada and Watson, Peter, "The Death Of Hitler", *Washington Post*, 7–6–03

Record Group 263: Records of the Central Intelligence Agency, Records of the Directorate of Operations. Analyses of the Name File of Heinrich Mueller

Schellenberger, Walter, *The Labyrinth, Memoirs*, Da Capo Press, 2000

Schramm, Percy Ernst, *Hitler The Man And The Military Leader*, Academy Chicago Publishers, 1981

Schroeder, Christa, *Er War Mein Chef*, second edition, Georg Mueller Verlag, Germany, 1985

Sweeting, C. G., *Hitler's Squadron, The Fuehrer's Personal Aircraft and Transport Unit, 1933–1945*, Brassey's Inc., 2001

Taylor, Blayne, *Guarding The Fuehrer*, Pictorial Histories Publishing Company, 1993, p. 242

Toland, John, *The Last Days*, A Bantam Book/Random House, Inc., 1967

Trevor-Roper, Hugh, *Final Entries 1945 The Diaries Of Joseph Goebbels*, G. P. Putnam's Sons, 1978

—— *Last Days Of Hitler*, third edition, *The Times*, London, 1946

Winters, Jeffry, *Served Straight Up*, Supplement of *Mechanical Engineering Magazine* (100 years of flight), ASME, Dec. 2003, p. 20

Ziemke, Earl F., *Stalingrad To Berlin, The German Defeat in the East*, Center of Military History, U.S. Army, Washington, DC, 1968, p. 477

Lightning Source UK Ltd.
Milton Keynes UK
30 October 2009

145606UK00001B/40/A